670.

2.0⁹

CORNWALL COLLEGE
LEARNING CENTRE

User's Guide to

Rapid Prototyping

User's Guide to
Rapid Prototyping

Todd Grimm

Society of
Manufacturing
Engineers

Rapid Prototyping
Association
of SME

Dearborn, Michigan

Library of Congress Catalog Card Number: 2003114160

International Standard Book Number: 0-87263-697-6

Additional copies may be obtained by contacting:
Society of Manufacturing Engineers
Customer Service
One SME Drive, P.O. Box 930
Dearborn, Michigan 48121
1-800-733-4763
www.sme.org

SME staff who participated in producing this book:

Rosemary Csizmadia, Editor/Production Supervisor
Frances Kania, Administrative Coordinator

Printed in the United States of America

Cover photos courtesy of:
3D Systems (wheel support and SLA time lapse)
Accelerated Technologies (valve)
Z Corporation (cell phone housings and Quincy compressor)
Ralph S. Alberts Co., Inc. (epoxy tool)

About The Society of Manufacturing Engineers (SME)

The Society of Manufacturing Engineers is the world's leading professional society supporting manufacturing education. Through its member programs, publications, expositions, and professional development resources, SME promotes an increased awareness of manufacturing engineering and helps keep manufacturing professionals up to date on leading trends and technologies. Headquartered in Michigan, SME influences more than half a million manufacturing engineers and executives annually. The Society has members in 70 countries and is supported by a network of hundreds of chapters worldwide. Visit us at www.sme.org.

About RPA/SME

The Rapid Prototyping Association of SME (RPA/SME) focuses on the technologies and processes that help conceive, develop, test, revise, and manufacture new products to cost-effectively bring them to market faster. Concurrent engineering and design for manufacturability are embodied in rapid prototyping technologies, which include fused deposition modeling, stereolithography, selective laser sintering, laminated object manufacturing, solid freeform fabrication, layered manufacturing, and others.

Rapid prototyping methods are timely and cost effective for large and small manufacturers alike. Beyond prototyping, however, rapid prototyping technologies are becoming production tools. Examples include producing consumable patterns for short-run casting or even rapid tooling for injection molding. In reverse engineering, a computer model created from scanning an object can be used to generate a physical copy of it.

Design engineers, product engineers, tool engineers, and manufacturing engineers interested in this fast-moving area are members of RPA/SME. They are in industries ranging from automotive to shipbuilding, and medical manufacturing to general service bureaus.

To my wife, Lori, with love

Without your support, patience, and understanding, this book would not have been possible.

Table of Contents

Appendix B: Additional Resources........ 351

Appendix C: Glossary.................... 367

Index 397

Preface

For the past 13 years, I have been a passionate supporter of rapid prototyping. As an advocate of the technology, I have presented at industry conferences, published articles and white papers, and participated on advisory boards. The fundamental motivation for these activities has been to create greater awareness and understanding of the rapid prototyping technologies. It is for this same reason that I have written *User's Guide to Rapid Prototyping.*

At times, the limited use of the technology has me baffled and frustrated. Having seen the powerful results that rapid prototyping can deliver, it amazes me that so few appreciate and experience these benefits. In an age of product development where better, faster, and cheaper is the mantra, how can anyone ignore the advantages of rapid prototyping and its ability to help companies achieve these three goals simultaneously?

At other times, when I am a bit more pragmatic, I understand why so few choose to implement rapid prototyping systems or use the services of a rapid prototyping bureau. There are a hundred reasons not to use this new (relatively speaking) technology. Without facts and proof of the advantages, it can be easy to downplay—or ignore—the benefits and highlight any disadvantages to avoid the risk and effort of change.

As you will see throughout this book, it is my belief, and the belief of many industry experts, that awareness, understanding, and education are the fundamental barriers to the growth of the industry. Of course, as with any tool, there are physical limitations that may be obstacles to its use. Yet, there is so little information available to help prospective users quantify the risks and limitations, and measure them against the benefits, to determine if rapid prototyping is a sound business decision. Lacking information, many find it easier to continue to use the tools with which they are comfortable.

As Terry Wohlers stated in the 2000 edition of the *Wohlers Report*, "the industry was in the chasm," a concept proposed by Geoffrey Moore in his book *Crossing the Chasm*. This is a period in the product life cycle where the aggressive, risk taking companies have adopted a technology, but the conservative, pragmatic types (the majority) have yet to be convinced of the benefits of investing hard-earned dollars and significant time into a relatively new, and yet to be fully proven, technology. I believe the industry continues to be restrained by the effects of the chasm.

I am a believer, a supporter, and an evangelist of rapid prototyping. My goals center on the fundamental belief that the rapid prototyping industry must help designers, engineers, executive management, and all others in the corporation to better understand the technology so that they can make informed business decisions. Is the technology right for every industry, business, product, and application? No. Is it worth considering? Absolutely.

I want to begin breaking down the communication and knowledge barriers with this book. It is only one small step towards the goal of rapid prototyping becoming a widely used tool. But hopefully, it is the first step in your organization's journey to truly understand rapid prototyping.

I hope that you enjoy the following discussion. And more importantly, I sincerely hope that I have contributed to a better understanding and appreciation of rapid prototyping.

Acknowledgments

This book would not have been possible without the support and contributions of many people and organizations in the rapid prototyping industry. I would like to thank each for their assistance and guidance.

In 1990, Laser Prototypes, Inc. (Denville, New Jersey), a rapid prototyping service bureau, asked me to join the company. With that invitation, I was introduced to the rapid prototyping industry. Over the next five years, I gained experience and knowledge about rapid prototyping technologies, applications and benefits. This information is woven into each chapter of the book. It also served as the foundation upon which future experiences were based. In 1995, I joined Accelerated Technologies (Austin, Texas), also a rapid prototyping service bureau. This company helped me to develop a greater understanding of rapid prototyping, and it allowed me the opportunity to share that information with the industry. Much of this book is a direct result of my experiences at Accelerated Technologies. Additionally, some of the data and many of the application stories come from this company's efforts and support.

I would also like to thank Terry Wohlers (Wohlers Associates, Fort Collins, Colorado), a business associate and friend. It is with his support, encouragement, guidance, and knowledge that I have reached a point in my career where I have the opportunity to write a book on rapid prototyping.

There are many others who were instrumental in the development of the book, including: Mike Durham (Accelerated Technologies), Bob Best and Joe Roth (Fisher Design), Elaine Hunt (Clemson University), Greg Turcovsky (Carlson), Dimitri Dimitrov (University of Stellenbosh), Lauren Groth (Ultimate Solutions, Inc.), Frank Leyshon (Leyshon Miller Industries), and Ed Grenda (Castle Island/ *Worldwide Guide to Rapid Prototyping*).

Contributors that I would like to thank include: 3D Molecular Designs; 3D Systems, Inc.; Armstrong Mold Corporation; Auburn Engineering; Bastech, Inc.; CAD/CAM Publishing; Center for BioMolecular Modeling; DSM Somos; Dynacept Corporation; Extrude Hone; D- M-E MoldFusion; General Pattern Company; LGM (Laser Graphic Manufacturing); Materialise; Medical Modeling, LLC; Ralph S. Alberts Co., Inc.; Roland DGA, Solidica, Inc.; Stratasys, Inc.; University of Louisville; and Z Corporation.

And finally, I would like to thank the Society of Manufacturing Engineers (SME). As the publisher, SME turned this book into a reality. As an organization, it has helped me, and countless others, to gain insight into rapid prototyping through its events, seminars, and publications.

CHAPTER 1

Introduction

Rapid prototyping is amazing, powerful, and revolutionary. Since the delivery of the first rapid prototyping system, the scope of applications and breadth of use have grown beyond belief. Virtually every industry that designs and manufactures mechanical components has used rapid prototyping. The technology is so pervasive that most people will use at least one product, on a daily basis, to which rapid prototyping has been applied.

Rapid prototyping is nearly a billion-dollar industry with more than 30 system vendors that have installed more than 9,500 machines around the globe (Wohlers 2003). With growth in the application of the technology for prototype development, other applications have come to light, namely rapid tooling and rapid manufacturing.

While the industry has grown, it is not without challenges. The general consensus is that less than 20% of the design and product development community uses rapid prototyping. In the manufacturing and manufacturing engineering disciplines, the level of use is far less.

If the technology is so powerful, why do so few companies use it? Why were early predictions—phenomenal, rapid growth and the replacement of conventional processes—never realized?

Rapid prototyping is a tool for design, engineering, and manufacturing. As with any tool, there are barriers and obstacles that impede its growth, and there are strengths and weaknesses that limit its use. It is amazing that prototypes can rise from a vat of resin or chamber of powder. It is powerful to produce parts without machining, molding, or casting. However, rapid prototyping is just a tool: an alternative solution to design and manufacturing challenges. The benefits and value of the technology are realized only when it is applied to suitable applications.

Determining when to apply rapid prototyping requires an understanding of the technology, the process, and its strengths and weaknesses. Industry leaders believe that this may be the key barrier to the rapid ascent of the technology. Many have concluded that a lack of awareness, understanding, and appreciation of rapid prototyping are critical barriers to its adoption and growth.

The goal of this introduction is to assist companies and individuals in assessing the merits of rapid prototyping and developing a full understanding of this unique and powerful tool. With this description, each can make an informed, personal decision regarding the applicability or necessity of rapid prototyping in the product development process.

CREATING UNDERSTANDING AND AWARENESS

Magazine articles, conferences, technical articles, books, and even television news programs have featured and discussed rapid prototyping. Nearly every trade show that serves the design and manufacturing communities has a rapid prototyping presence. There are thousands of web pages available on the Internet that promote and discuss the technology. Yet, there continues to be a scarcity of information that offers detailed analysis, review, and comparison.

Much of the publicly available information addresses the obvious advantages of rapid prototyping, focusing on the remarkable ability to grow parts from digital data. Much of what remains focuses on the latest systems and materials developments, often resulting from vendor-issued promotional materials. What is missing is information that details the true experiences of rapid prototyping users. Without an accurate account of both the advantages and limitations, from a user's perspective, the knowledge gap impedes an increase in awareness and understanding.

Gathering information to develop an understanding is especially difficult for those who are new to the rapid prototyping industry. As an introduction to rapid prototyping, the discussion of the technology targets the majority of those in industry, those who have yet to apply it. Yet, this is not a light review of the technology. The detailed accounts and user insights offer a deep appreciation for the technology and the process. Even experienced rapid prototyping users will find valuable insight and information.

To fill the knowledge gap, this introduction to rapid prototyping details the process, individual technologies, applications, strengths, and limitations. It also offers a comparison of rapid prototyping technologies with processes like machining.

WHAT IS RAPID PROTOTYPING?

One benefit of a rapid prototype is that it improves communication. However, the technology is often a source of miscommunication and misunderstanding.

There are numerous terms for rapid prototyping, including: freeform fabrication, solid freeform fabrication, autofab, automated freeform fabrication, digital fabrication, 3D printing, laser prototyping, layer-based manufacturing, additive manufacturing, and solid imaging. The multitude of terms and definitions can confuse a discussion or description of rapid prototyping.

Equally confusing is that a simple term like 3D printing has multiple definitions. Some in the rapid prototyping industry use the term 3D printing to characterize all varieties of rapid prototyping technology. Others apply the term to a specific class of rapid prototyping systems. With the great disparity in definitions and terms, it is critical that there is an agreement and understanding in any discussion of rapid prototyping. Without this agreement, miscommunication is likely.

This is Not a Book About Prototyping

Within the design and manufacturing communities, other factors contribute to the confusion. There is disagreement on the technologies that should be included under the umbrella of rapid prototyping. Many suppliers of technologies and materials for processes as varied as machining and molding promote their offerings as rapid prototyping. While each truly prototypes rapidly, they are subtractive or formative processes. To include them in the discussion that ensues would require an introduction to prototyping, a topic that is much too broad for a single book.

This is a Book on Rapid Prototyping

Instead, this is a book dedicated to rapid prototyping. This book is about additive processes that eliminate machining, tooling, molding, casting, and fabrication. The election to address only additive technologies is not an indication that other processes are not rapid. As will be illustrated, rapid prototyping may be a slower process or a weaker solution for a project. Combined with the extensive information and first-hand experience available for other, conventional processes, this detailed account of rapid prototyping promotes the ability to select the best technology and apply it wisely.

The technology decision is personal and unique. In some cases, rapid prototyping will be the best. However, as will be shown, in most situations the selection of the best tool will not be obvious. With ample amounts of crossover, both rapid prototyping and the competitive technologies are likely to serve a user's needs. These conventional processes are often rapid prototyping solutions; however, they are not additive rapid prototyping solutions and, therefore, they will not be discussed.

DETAILED TECHNOLOGY DESCRIPTION

This introduction to rapid prototyping will cover, in detail, all aspects that are important for a clear understanding and appreciation of the technology. Chapter 2 begins with an overview of the technology, its applications, and benefits. In this overview, rapid prototyping is clearly defined.

While rapid prototyping can be a push button, one-hour process, much more goes into most prototypes. To develop an understanding and to build a foundation of information on which to build, Chapter 3 details the rapid prototyping process. Through this description, there will be a greater appreciation for what it takes to successfully build rapid prototypes and the information required for making good technology decisions.

For further clarification, classes of rapid prototyping are presented in Chapter 4. While the rapid prototyping industry lacks consensus, this proposed classification system helps to distinguish the differences between a $30,000 and an $800,000 system. Although the methodologies and output are similar, there is a great variance in operational demands, user control, and final results.

Chapter 5 discusses applications and benefits. As indicated, rapid prototyping is more than a prototyping tool. Its applications cover the full spectrum of design and engineering and extend to applications in manufacturing. Additionally, there are examples of applications outside of the confines of design and manufacturing. From these examples, new ideas and unique, innovative solutions may come to mind.

Since some include machining in the rapid prototyping application set, and since machining can be as fast or faster, a detailed comparison of rapid prototyping and computer numerical control (CNC) machining is provided in Chapter 6. With this head-to-head comparison, it will be possible to determine when and how to apply each tool.

Using four leading technologies, Chapter 7 provides a head-to-head comparison of rapid prototyping systems. Containing information known only to rapid prototyping users, this comparison reveals key considerations of the technologies, exposes some little known truths, and eliminates common misperceptions.

For those who find that rapid prototyping could be valuable, the path through justification, evaluation, and implementation may be challenging. Chapters 8 and 9 are guides for the selection and implementation process.

Although the focus is on prototyping, an introduction to rapid prototyping would be incomplete without a discussion of rapid tooling and rapid manufacturing. These applications are intertwined with rapid prototyping, and it would be inappropriate to exclude them. In addition, many believe that these two applications may be the areas of significant growth in the coming years. Chapter 10 addresses both rapid tooling and rapid manufacturing.

To complete the introduction to rapid prototyping, Chapter 11 summarizes the key aspects of rapid prototyping and forecasts the role that the technology will play in the future. Whether or not the decision is to use rapid prototyping, one must keep abreast of the coming changes.

Appendices A through C offer supporting information. Appendix A provides user case studies that show the real-world benefits of rapid prototyping. Appendix B lists useful resources that may be helpful in the further evaluation of the technology. And finally, a glossary of terms is provided in Appendix C.

SHADES OF GRAY

There is not an answer that is right for everyone. The benefits of rapid prototyping are always in question until the question is asked in the context of specific needs and goals. If rapid prototyping is a viable tool, the answer to the question of which technology is best will be unique, personal, and individual. Therefore, the information is not delivered with definitive statements. Instead, the technology is discussed in a way that allows the reader to develop his or her own answers.

This book has something for everyone. For those who believe in rapid prototyping, there is information that will support their convictions. For those who want to prove that rapid prototyping is an inappropriate solution, there is plenty of justification. And for those who simply want to determine the right answer, there is plenty of information to aid in the decision-making process. How can all of these desires be satisfied at once? It is simple. This journey through rapid prototyping offers no definitive statements, and in many cases, it is delivered in a way that invites more questions. The information offered carries an overriding principle that the answer to each question is "it depends."

The strengths and weaknesses, applications and benefits, and evaluation and implementation of rapid prototyping are unique to each part, product, program, and company. Therefore, the answer to every question will be "it depends." The answers can only be determined when the information on rapid prototyping is combined with the specific and unique circumstances within each company and for each project.

A TOOL FOR CHANGE

Faced with economic challenges and global competition, the way business is done is changing. Organizations around the globe need to drive costs out of the process and product while enhancing quality and reducing time to market. Those who shoulder the burden of these requirements and initiatives find themselves with more work, fewer resources, and crushing deadlines. To cope or excel in this environment, the way business is done has to change.

Although this change will come in many forms and take years to develop, two key elements are collaboration and innovation. Design engineering and manufacturing engineering need to eliminate the barriers between the departments. Rather than "throwing a design over the wall," design and manufacturing should communicate early in the process. This communication will produce a better product at less cost and in less time. To innovate, change is required, and this change demands that nothing is taken for granted and that no process is sacred. New methods, processes, and procedures are required in the highly competitive business environment.

Rapid prototyping may be the tool for change. To realize its full potential, rapid prototyping should be adopted by all functions within an organization. If implemented, it should not be designated as merely a design tool. Manufacturing needs to find ways to benefit from the technology, and it should demand access to this tool. This is also true for all other departments: operations, sales, marketing, and even executive management. When adopted throughout the organization, rapid prototyping can be a catalyst to powerful and lasting change.

INFORMED DECISIONS

The rapid prototyping industry has achieved much since its inception. There have been major advances since the 1987 introduction of stereolithography. Yet, there is room for growth and a need for further advancement.

Many of the obstacles that rapid prototyping faces are not unique. As with any new technology, there is a resistance to change and a reluctance to work through the challenges of a developing technology. However, there are other factors that are unique to this industry. Since rapid prototyping requires 3D digital definition of the part, its growth rate is the same as that of CAD solid modeling, an application that is far from being used by the majority of design professionals. Additionally, rapid prototyping has been burdened with a negative perception that the parts are "brittle." While this was true many years ago, this is no longer an appropriate generalization. Yet, many use the belief that rapid prototypes are brittle to justify not evaluating or using the technology.

Since it continues to be a tool for the minority, rapid prototyping may not pose a competitive threat to those who do not use it. However, many companies that have implemented the technology have discovered powerful advantages in applications that range from product development to manufacturing to sales and marketing.

The decision is yours. This introduction to rapid prototyping offers no answers. Instead, it offers information to assist in making an informed decision.

REFERENCE

Wohlers, T. 2003. *Wohlers Report 2003: Rapid Prototyping and Tooling State of the Industry Annual Worldwide Progress Report.* Fort Collins, CO: Wohlers Associates, Inc.

CHAPTER 2

Overview

In 1987 an industry was born. In that year, 3D Systems unveiled the world's first rapid prototyping device. 3D Systems' stereolithography technology ushered in an age of direct, digital technologies for the rapid production of models, prototypes, and patterns.

Rapid prototyping is an exciting and powerful technology. It has changed the design, engineering, and manufacturing processes in industries as diverse as consumer products and aerospace. Originally addressing the delays in prototype construction, every aspect of the design process now uses rapid prototyping. It has also broadened its application scope with solutions for tooling and manufacturing.

DEFINITION OF RAPID PROTOTYPING

"Rapid prototyping" has become a generic term used to describe many prototyping processes, as shown in *Figure 2-1*. It is regretful that the industry was built upon a simple, powerful, and attractive phrase. Today, everything is about speed, efficiency, and productivity, so the words are commonly applied to any process that generates prototypes rapidly. In addition, "rapid" is a relative term, so it is applied both to processes that build prototypes overnight and to those that take a week or more. As long as the process is completed faster than was previously possible, companies and industries are applying the rapid prototyping label.

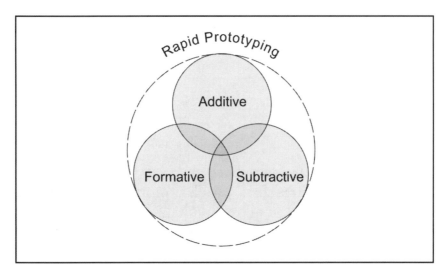

FIGURE 2-1. Rapid prototyping has become a generic term for all processes that generate prototypes quickly.

Without a definition of rapid prototyping, it would be impossible to address the topic clearly and concisely. In support of this statement, observe any rapid prototyping conversation. In the majority of instances, it will become clear that there is an inconsistent definition of the technology. As the conversation progresses, the communication barrier grows until someone asks the question, "What do you mean by rapid prototyping?" To further support this point, simply do a search in any Web search engine. "Rapid prototyping" yielded 250,000 results in Google. While the first three pages of results were right on target, they quickly degraded to links for molding, machining, casting, and tooling. Deeper into the search results, rapid prototyping took on an entirely different meaning since it is a common term used in the printed circuit board, software development, audio track, and hard disk array industries.

For this text, rapid prototyping is used in the context of its original meaning, which was coined with the commercial release of the first stereolithography system.

Rapid prototyping:　A collection of technologies that are driven by CAD data to produce physical models and parts through an additive process.

In even simpler terms, rapid prototyping is a digital tool that grows parts on a layer-by-layer basis without machining, molding, or casting, as shown in *Figure 2-2*.

To an even greater extent, rapid tooling and rapid manufacturing are subject to multiple definitions. Once again, if the process is completed quickly, many will describe it as either rapid tooling or rapid manufacturing. For this text, the definitions of these technologies are:

Rapid tooling:　The production of tools, molds, or dies—directly or indirectly—from a rapid prototyping technology.

Rapid manufacturing: The production of end-use parts— directly or indirectly—from a rapid prototyping technology.

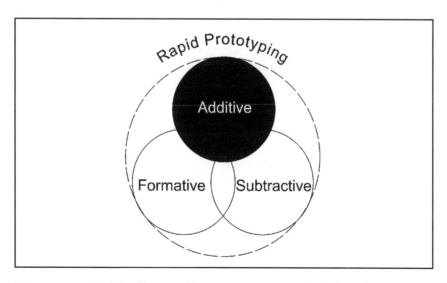

FIGURE 2-2. Originally, rapid prototyping included only additive processes. This is the area of focus for this book.

Direct processes are those that produce the actual tool (or tool insert) or sellable part on the rapid prototyping system. Indirect processes are those where there is a secondary process between the output of the rapid prototyping system and the final tool or sellable part.

HISTORY OF RAPID PROTOTYPING

It is hard to capture the enthusiasm, fascination, and awe of the early days of rapid prototyping in words. The excitement, anxiety, challenges, and fears in the first years of the technology cannot be fully expressed. To truly appreciate and understand the early years, it had to be experienced.

With the public announcement of the amazing device that transforms digital data into real parts, the media was struck with the same fascination as those in the mechanical engineering and computer-aided design (CAD) communities. Television programs like "Good Morning America" and "Wall Street Week" introduced the technology to an awestruck audience. Articles began popping up in both trade journals and mainstream news magazines.

With the fanfare and excitement created by the announcements, many predicted overnight growth to a billion-dollar industry. Others forecasted the doom of established technologies like numerical control (NC) and computer numerical control (CNC) machining. However, the pace of growth was, and continues to be, slower than expected. After more than a decade, rapid prototyping is a tool used by the minority. After more than 10 years, rapid prototyping faces many of the same struggles and challenges that existed in the beginning.

Timeline

On March 11, 1986, Charles (Chuck) Hull received patent number 4575330 (see *Figure 2-3*). The patent, *Apparatus for Production of Three-Dimensional Objects by StereoLithography*, represents the birth of the rapid prototyping industry (Hull 1986). Although others were also pursuing the dream of printing prototypes, the industry recognizes Chuck Hull as the father of rapid prototyping.

In that same year, 3D Systems was founded. One year later, 3D Systems introduced the first commercial rapid prototyping device, the SLA®-1, which is pictured in *Figure*

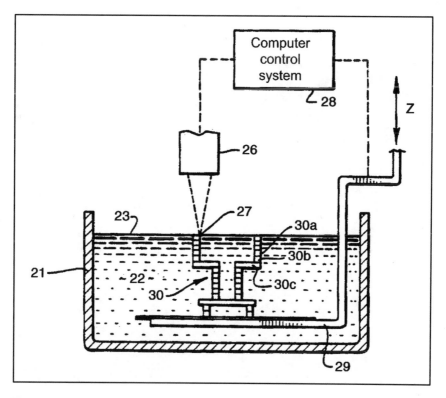

FIGURE 2-3. U.S. patent 4575330, issued to Chuck Hull, marked the birth of the rapid prototyping industry. *(Courtesy U.S. Patent and Trademark Office)*

2-4, at AUTOFACT in Detroit, Michigan. After completing a beta program, which included companies such as AMP, General Motors, Pratt & Whitney, Baxter Healthcare, and Eastman Kodak, the first commercial SLA-1 was sold in 1988. With an upgraded recoating system, the SLA-1 became the SLA 250, which was released in 1989. A year later, the larger SLA 500 became available (Jacobs 1992).

Almost as soon as the SLA-1 became available, service bureaus began to pop up in the United States and around the world. By early 1988, three service bureaus were in operation: Plynetics, Swiss Wire, and Laser Prototypes. These pioneers were soon followed by many companies. The early success of the service bureaus was built upon

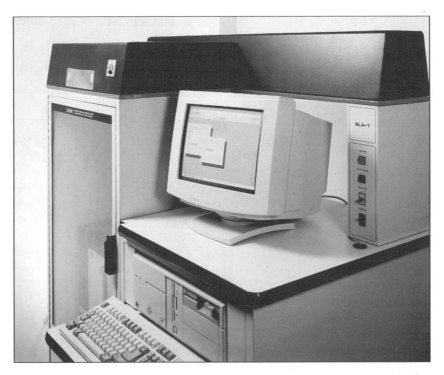

FIGURE 2-4. The first commercial rapid prototyping device, an SLA®-1 with serial number 880001 from 3D Systems, continues to produce parts at Accelerated Technologies. *(Courtesy Accelerated Technologies)*

offering access to rapid prototyping without the high cost of a system purchase and the risk of investing in a new, developmental technology.

Expectedly, 3D Systems was not alone for long. Between 1988 and 1992, many competitive technologies were released, and many others were in development. The lure and attraction beckoned companies and individuals to join the rapid prototyping industry. Companies such as Ballistic Particle Manufacturing, CMET, Quadrax, Cubital, Helisys, Stratasys, EOS, and DTM had started their businesses or released their technologies (Wohlers 2003a). Some were similar to photopolymer-based stereolithography, while others had dramatically different approaches. In this timeframe, processes based on laser cutting, lamination, extrusion, sintering, and jetting were born. The timeline in *Figure 2-5* shows some of the early developments.

By 1990, rapid prototyping conferences had emerged. In that year, the Solid Freeform Fabrication Conference (Austin, Texas) and the International Conference on Rapid Prototyping (Dayton, Ohio) provided attendees from around the world an opportunity to learn about this new technology.

In 1991, the first commercial offering to stem from the Massachusetts Institute of Technology (MIT) research and development of jetting was launched. Soligen, the license for MIT's technology as it applies to jetting of a binder onto ceramic, launched its mold-building service for investment casting. Over the years, the MIT patents for its 3DP process would give birth to other companies and technologies, including Z Corporation and Extrude Hone's ProMetal® system. The first successful attempt at patent protection came in 1992, which was the beginning of what has become common practice. Quadrax, after selling just two systems, closed its doors. Patent litigation on the part of 3D Systems forced Quadrax to cease operations.

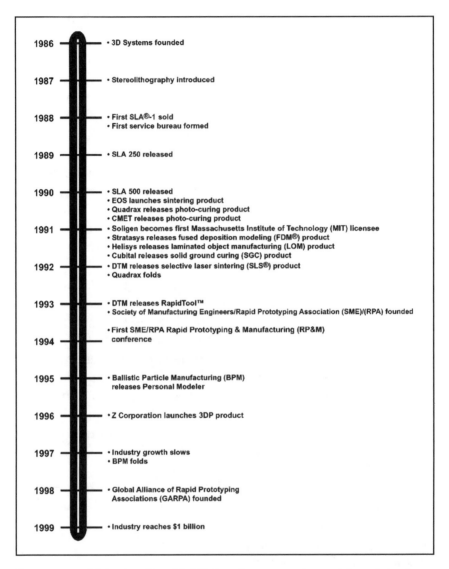

FIGURE 2-5. This timeline highlights the events in rapid prototyping.

In 1993, the term "rapid tooling" began to take hold as both a process and a goal. In that year DTM announced its RapidTool™ product, a system that delivered sintered metal tooling inserts for injection molding and other manufacturing applications. This came on the heels of the realization that to address only the development of a single prototype was not as useful as delivering a mold or pattern to be used in other processes. In the early 1990s, companies developed processes that used rapid prototypes as patterns for rubber molding, epoxy tooling, investment casting, and other conventional molding operations. The term "rapid tooling" caught on because many desired the ability to deliver injection molds quickly. To this day, the term and concept are popular, and there continues to be a struggle to exceed the capabilities of conventional technologies like CNC machining.

Recognizing RP as an industry and application that was viable with great potential, the Society of Manufacturing Engineers (SME) created the Rapid Prototyping Association of the Society of Manufacturing Engineers (RPA/SME) in 1993. RPA/SME's origin is 1992 when Terry Wohlers spearheaded the creation of the first rapid prototyping association, which became the RPA/SME.

The first RPA/SME Rapid Prototyping & Manufacturing (RP&M) conference and trade show was held in 1993. Although SME had sponsored rapid prototyping related events since 1989, this was the first official show for the Association. The event brought together a small gathering of like-minded professionals involved in, or thinking about, rapid prototyping. Today, this annual event hosts dozens of speakers, hundreds of exhibitors, and thousands of attendees.

The shift from small, entrepreneurial service bureaus to larger, full service businesses also began in 1993. In prior years, most service bureaus possessed a single machine, and those that owned two were exceptional. With the sale of DTM's service bureau operations to Accelerated Technologies, Inc., service bureaus became serious business. With this purchase, Accelerated Technologies became, by far, the largest service bureau in the world, operating eight selective-laser-sintering machines. This move ignited the

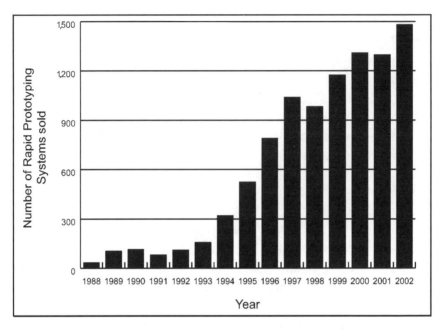

FIGURE **2-6.** Global rapid prototyping system sales have risen steadily, as shown by the number of systems sold by year (Wohlers 2003b).

growth of service bureau operations. Companies like Plynetics, Prototype Express, and Compression Engineering followed Accelerated Technologies' lead to tackle growth. That year also marked the shift to full service operations that extended beyond rapid prototyping. Regrettably, rapid growth led to the demise of some operations. After the merger of Plynetics, Prototype Express, and Laserform into Plynetics Express, the company closed its doors. Soon after, Compression Engineering followed.

In 1996, Z Corporation began selling its 3D printing systems. Based on a license from MIT, the technology broke two barriers: speed and cost. This combination was the recipe for success for this company. By addressing the user demands of faster and cheaper rapid prototyping technology, the company also spurred the development of a new industry segment, 3D printers. In later years, other

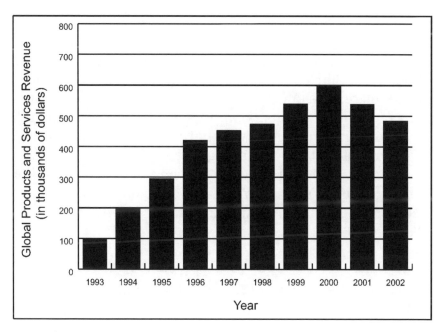

FIGURE 2-7. Global product and service revenues have been inconsistent since 1997 (Wohlers 2003b).

companies like Stratasys, Objet Geometries, and 3D Systems would follow suit.

Throughout the 1990s, rapid prototyping associations around the globe were formed. To unify these international efforts, Terry Wohlers and Dr. Ian Gibson led the formation of the Global Alliance of Rapid Prototyping Associations (GARPA). The first GARPA meeting was held in 1998 in Dearborn, Michigan. Today associations from 15 countries are members of GARPA.

Throughout the period from 1988 to 1997, the industry had tremendous growth, averaging 57% per year. *Figure 2-6* shows the number of rapid prototyping systems sold globally by year (Wohlers 2003b). Adoption of the technology, addition of multiple systems and technologies, and growth of the service bureau market fueled the growth. After 1997, the results were spotty. Although some years were better than others, for the most part, there was only minor growth from year to year, as illustrated in *Figure 2-7*, which shows product and service revenues by year (Wohlers 2003b).The reasons are varied and numerous. The role of the service bureau was diminishing, used equipment was readily available, and supply had outstripped demand. However, the key factor was that the majority had yet to accept and adopt the technology.

Currently, rapid prototyping is nearly a billion-dollar industry with more than 30 system vendors that have installed more than 9,500 machines around the globe (Wohlers 2003b).

Life in the Early Days of Rapid Prototyping

It was 1987 and I was selling computer-aided design/computer-automated manufacturing (CAD/CAM) systems for McDonnell Douglas. This new, inexpensive tool called AutoCAD® was poaching a lot of our low-end business. Competition was fierce, and I was looking for a better way to differentiate my product.

I opened an industry magazine, probably *Machine Design* or *Mechanical Engineering* (I cannot remember which, but that is not important). In the news section, I saw a picture of an odd looking device that claimed to produce prototypes directly from CAD data without machining or molding. I was awestruck. This was the answer. This was the future. That picture showed 3D System's SLA-1.

Three years later, Laser Prototypes, Inc., one of the first rapid prototyping service bureaus in the world, invited me to join the company. With a single SLA 250, Laser Prototypes was a world leader. At the same time, 3D Systems began running their "overnight prototypes" ad, which created many obstacles to growth. For at that time, less than 1% of all prototypes could be created overnight. Believing what they read, customers asked for their prototypes overnight, but we would inform them that delivery would be five to seven days. While this was lightning fast when compared to conventional model making, the customers expectations were set too high, and when these expectations were not met, they often turned their backs on the technology.

So, why did it take five to seven days for even the most reasonably sized prototype? Technology. In those days, most people did not even use 3D CAD, and those who did often did not have the ability to generate a stereolithography (STL) file. (STL is an ASCII or binary file format used as input for rapid prototyping devices.) So, painstakingly, the 2D data was converted to 3D models and then to an STL file. But that was not the end of the technological challenges.

With the STL in hand, the data needed to be prepared for building. Using a text-based interface on a 10 MHz, 20 MB hard drive, DOS computer, the files were laid out without any visual feedback. There was only one way to see if the build was laid out correctly, and that was by previewing it slice-by-slice on the SLA 250. This was after completing the setup and slice, which would take as long as 12–48 hours! Once we had a good build file, success was still elusive.

Building parts was a crapshoot. You never knew what would come out of the machine. The term "trapped volume" had not even been coined yet, let alone was there a solution for it. (Trapped volumes are pockets in a prototype that, in the stereolithography process, hold resin. In early systems, trapped volumes would create a dome of material that rose above the working surface of the liquid polymer.) In those days, the 30 mW systems with slow photo-speed resins meant that builds commonly took 8–48 hours. I can still remember arriving to work praying that the build from the previous day had not crashed. But often it had, and it was time to start all over.

When a build was successful, the job still was not complete. Inaccuracy, poor surface finish, and missing features often caused us to go back to the beginning. Even if the part came out reasonably well, it was time to take on the challenge of support removal. The original material, 5081, deserved its reputation as brittle. Sometimes it seemed that even a slight breeze caused the part to break. When applying files, Dremel® tools, and sandpaper to the prototype, you were inviting trouble. I would say that most prototypes in those days had some repair work done to them to reattach broken features.

After all of this, I would still hold my breath in anticipation. You never knew if the client would be satisfied with the results or if FedEx would deliver a package with undamaged contents. Two situations come to mind. One involves a class ring and the other drug-sniffing dogs. After days of prepping, building, and finishing a model for Grumman Aircraft, a beautiful prototype was hand delivered to the conference room where top management awaited their part. It was beautiful, a real source of pride. As Bill Coleman, a founder of Laser Prototypes, discussed the prototype that was placed prominently in the center of the conference table, he made a hand gesture towards the model. To his horror, his college ring caught the corner of the prototype, and being brittle, a large chunk broke off and fell to the table. The other story involves Motorola's Florida operations. Once again, Laser Prototypes painstakingly produced a beautiful rapid prototype. It was carefully packaged and overnighted to Motorola. The next morning we called, expecting delight from our customer. To our dismay, the client told us that the prototype arrived in pieces. We rebuilt and resent the prototype, only to have the same thing happen. Confused, we began to investigate. What we found is that the drug-sniffing dogs at the airport had barked at both packages. So, officials opened them for inspection. However, they did not appreciate the fragile nature of these models, and they just threw them back in the box. Unprotected, the prototypes were broken on the last leg of their journey.

Even with the challenges, anxiety, and delays, it was exciting and rewarding to be involved in this brand new industry. I can remember delivering prototypes to first-time rapid prototyping users. These talented professionals would often run around their office, like little kids, showing off their prize.

Today, all of these issues have been addressed. Relatively speaking, times are good. Yet, as users, we want more. We continue to push the technology to be better, faster, and cheaper. The good news is that the industry continues to respond.

WHY PROTOTYPE?

Prototype, prō'tō-tīp *n*. An original or model; a pattern. Derived from Greek proto-typon, meaning first form.

Prototyping is a tool to improve communication. The many benefits of prototyping stem from this basic value offered by prototypes of all kinds. A prototype clearly communicates a new product's design to everyone in the design, manufacturing, and decision-making processes. Whether this communication is feedback to the design engineer or a selling tool to prospective clients, the prototype serves to simply, accurately, and quickly communicate the new product's design in a form that makes the intangible concept physical and real. Prototypes of all types, whether rapid or conventional, share this fundamental advantage.

"Nothing is so simple that it cannot be misunderstood."
J. R. Teague

For some, prototyping is perceived as a discretionary tool that is only used when time and budget allow. For others, perhaps the wisest, prototypes are employed for even the simplest and routine task. If a prototype demonstrates that a design is perfect, it may not serve any purpose and may be perceived as an unnecessary expense. However, the worst case is when a prototype is not developed and a problem makes it all the way to manufacturing or to the customer's purchase. This scenario teaches the design community that it is unwise to move forward without the validation offered from the prototype.

Over the years, prototyping has been likened to an insurance policy. Until there is a claim, there is no benefit or value. Some choose to forgo the insurance policy to save money. These individuals are betting that there will be no need for, and no claim against, the policy. Others elect to have the protection and peace-of-mind the insurance policy offers. Prototyping is an insurance policy against defect and errors.

When a design takes physical form, ambiguity, assumptions, and perceptions are removed from the equation. By eliminating the need to interpret engineering drawings or 3D CAD data, all those involved in the development and launch of the new product can understand, whether or not they are technically oriented.

"Never before have we had so little time in which to do so much."

Franklin D. Roosevelt

Product designs are far from simple. If even simple things can be misunderstood, how can the complexities of these designs possibly be communicated without misunderstanding? This design complexity is compounded by the constant drive to produce it faster and cheaper. The corporate initiatives of time-to-market reduction, quality improvement, and cost reduction can create an unmanageable environment for the design professional. Prototypes aid the corporation in achieving its goals while assisting the design team to cope with the challenges of too much to do, too little time, and too few resources.

A Toolmaker's Nightmare

A client ordered injection-molded parts of an old design that included only a few minor modifications. The toolmaker/molder proposed the construction of a prototype prior to committing to tooling. But the client resisted, since the company had been doing the part for years.

From CAD data, the tool was constructed, parts were molded, inspection was performed, and the components were delivered. The work was performed flawlessly, on time and on budget. However, upon contacting the client to confirm the delivery and acceptance of the parts, the molder found that he had an angry client. Why? The parts were twice the size that they should have been.

After a thorough investigation, the molder determined that the CAD data he received was scaled to twice its nominal size. It appeared that one of the client's employees had scaled the data and mistakenly saved it.

If the client had elected to receive a prototype, the error would have been detected prior to investing time and money in useless tooling.

While it is rewarding and uplifting to produce the perfect prototype, one that shows the world a flawless design, this is not the goal of prototyping. Rather the goal is to uncover errors, mistakes, and flaws prior to a design's release to manufacturing. The true goal of prototyping is not to prove that the design is correct but rather to reveal any detail that will affect cost, quality, time, or consumer acceptance. Prototyping is about uncovering mistakes. And every design project is loaded with them since, as Theodore Roosevelt said, "The only man who never makes mistakes is the man who never does anything."

BENEFITS

"Prototype early and often."
Curtis Bailey, President,
Sundberg-Ferar

All prototypes offer the fundamental advantage of improved communication for the detection and correction of errors and flaws, and rapid prototyping shares this critical benefit. The differences between rapid prototyping and conventional methods arise from the process. Of course, the key element of the process is the physical construction of the prototype in an additive fashion. But to understand the breadth of rapid prototyping's benefits, the entire process must be considered.

In its simplest form, the key advantage of rapid prototyping is that it delivers models quickly. With a name like "rapid prototyping," this is obvious. What is hidden is that rapid prototyping is not always the fastest method. Rather, rapid prototyping is fastest for prototypes with complex geometry. The time advantage swells as details, features, and freeform surfaces are added to the model. The reason is that, due to the additive nature of the technology, rapid prototyping is virtually insensitive to any level of design complexity. Design features like undercuts demand more setups, time, and consideration in machining; that is if they can be produced directly in the model. Undercuts also add challenges to all forms of tooling. But for rapid prototyping, the undercut is easily accommodated without any time or

FIGURE 2-8. The caged ball, constructed on a stereolithography machine in 1992, was an early tool to demonstrate the advantages of rapid prototyping.

cost penalty. The ability to construct complex geometry is illustrated in *Figure 2-8*, which shows a circa-1992 caged ball constructed by stereolithography in one operation.

Rapid prototyping can reproduce—quickly—designs that are unthinkable by any other method. For example, in rapid prototyping a sphere can be constructed within a sphere in just one operation, as illustrated in *Figure 2-9*. In fact, the overall machine time and expense is actually less to build one within the other than to build two separate spheres. In conventional processes, this design would be fabricated in pieces, and later assembled, to allow a cutting tool or mold surface to carve or form the spherical shapes. *Figure 2-10* shows a working assembly constructed in one build.

The time advantage does not begin and end with the physical part construction. In fact, the entire process,

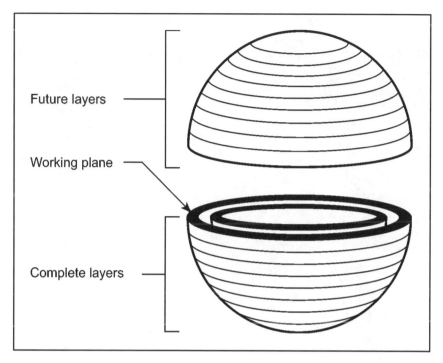

FIGURE 2-9. With the additive process, rapid prototyping can construct a sphere within a sphere, since the working surface (highlighted in black) is always the top of the last solidified layer.

which is highly automated, is faster than other methods. It would be feasible to receive data for the "sphere within a sphere" at 4:30 p.m. and deliver the model the next morning. This time advantage results from the direct processing of the electronic data in very few steps. Unlike machining, where the cutting strategy and required setups are considered prior to investing time in generating the tool path, rapid prototyping requires less analysis, fewer processes, and little labor.

Convenience and ease of use are also factors. Rapid prototyping offers a "path of least resistance," at least for the consumer of the prototype (maybe not for the operator). With few exceptions, the digitally driven nature of rapid prototyping is a direct route. An STL file is exported from a

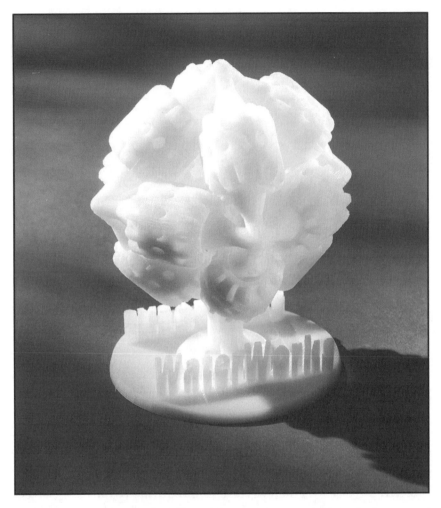

FIGURE 2-10. The "brain gear," constructed with fused deposition modeling, is a working assembly constructed in one operation. *(Courtesy Stratasys, Inc.)*

3D CAD file and turned over to the rapid prototyping operation. As long as the file is error-free, the process is as simple as specifying the desired technology and material. With conventional processes, there are many points where human intervention is required. At each of these points, assumptions may be made, decisions may be reached, or questions may arise. Each assumption, decision, or question can have an impact on the quality of the prototype, the time to deliver it, and the overall cost. So, when requesting a machined or molded prototype, time and effort, on the part of the prototype consumer, may be required to assure that desired results are achieved. This is not the case with rapid prototyping.

Often prototypes are excluded from the product development process to save time. Disrupting and delaying the design and analysis flow, while waiting on the physical embodiment of the design, is often justification to assume the inherent risk that comes with the decision not to prototype. With its overall speed and simplicity of application, rapid prototyping removes this barrier and offers not only error detection but also protection of the most valuable asset in business today, time. The time benefit applies to both corporate time to market and personal time management for the designer or engineer.

Of course, there are many other advantages, as well as limitations, to consider when evaluating rapid prototyping. However, the major advantages are time and efficiency.

OVERVIEW OF APPLICATIONS

Rapid prototyping's impact reaches far and wide. There is diversity in the application of rapid prototyping in terms of the disciplines that use it, the processes that benefit from it, the industries that employ it, and the products that are better because of it. The common element of all of these applications is that rapid prototyping has been a tool that makes the process faster, the product better, and the cost lower.

Industrial design, engineering, manufacturing, sales and marketing are just some of the disciplines that have applied rapid prototyping. The processes to which each has applied rapid prototyping match the breadth of these disciplines. A small sampling includes conceptualization, form, fit, function, tooling patterns, tool design, tool building, sales presentations, and marketing materials.

Every industry that makes metal or plastic parts has used rapid prototyping. Aerospace, automotive, consumer products, electronics, toys, power tools, industrial goods, and durable goods are some of the commonly referenced industries. With each successful application, the list grows. The technology is now applied to medical modeling, biomedical development, orthodontics, and custom jewelry manufacturing. Rapid prototyping is so pervasive that it would be unlikely any individual could go about his daily routine without using a product that has in some way benefited from rapid prototyping.

The list of products where rapid prototyping has been applied is too long to attempt to capture in a few words. Yet, some of the most exciting are those where rapid prototypes have actually taken flight in fighter aircraft and space vehicles. Equally impressive is that rapid prototyping is used to win races in NASCAR and Formula 1 car racing. Finally, rapid prototyping has even been used as a pre-surgical planning tool for the separation of conjoined twins.

Rapid prototyping is actually a misnomer when the entire breadth of applications is considered. With advances in rapid tooling and rapid manufacturing, rapid prototypes are no longer constrained to models, nor are they constrained to product development.

The following sections are just a small sampling of the potential applications of rapid prototyping. As more research and innovation is directed to this technology, more, and possibly unthinkable, applications will arise.

Product Development

Following are the possible applications of rapid prototyping in product development.

- Concept models are short-lived models for early design
  validation or communication. This application could
  also be called a visual aid. Much like a printed first draft
  of a manuscript, the concept model is used for proofing
  and quickly discarded. Multiple design iterations are
  often reflected through concept models (see *Figure
  2-11*).

- After the concept has been defined, the design
  progresses into its mechanical definition. At this stage,
  the prototype is used to check form and fit, confirming
  that parts properly mate and there is ample space
  allowed for components. The other aspect is confirmation of the overall design form and its suitability for the
  intended use.

- Closely related to form checking, ergonomic studies
  take the review one step further by analyzing the interaction of the product with the user. Here, elements such
  as balance, grip, accessibility and overall comfort are
  reviewed.

- Whether a design will work in the intended application
  is the question that functional testing answers. Here the
  prototype is subjected to external forces, environmental,
  mechanical, and electrical, in a mock operating environment to determine if the product will work as
  intended without component failure.

- Components are sent out for a request for quote (RFQ) to
  solicit bids for the manufacturing process. Traditionally,
  engineering drawings (blueprints) were issued with the
  request, and later, CAD data was issued. Studies have
  found that the submission of a physical prototype with
  the RFQ can dramatically reduce the vendor's price.

FIGURE 2-11. Concept models of this cell phone housing offer rapid design review and iteration.

- In proposals and presentations, the use of a prototype as a communication device is beneficial, especially when communication is to those who do not understand prints or CAD. In this application, the prototype is dressed (painted and decorated) to look like the end product. These dressed models are used in tradeshows, sales meetings, and photography shoots to allow customers to fully appreciate a product prior to its manufacture. They are also used in internal meetings, such as sales meetings, to aid staff in understanding the new product.

- Since rapid prototyping requires no manipulation or human interpretation of the design data, its direct output offers verification of the quality of the CAD model. A corrupt feature or small defect in the CAD file can prove disruptive to downstream manufacturing processes. Since rapid prototyping is a direct translation of the CAD file, any errors or corruptions can be detected and corrected prior to freezing the design (see *Figure 2-12*).
- While collaboration has gained in popularity, engineering and manufacturing continue to be two disciplines that are separated, at least in many shops. What an engineer designs has tremendous impact on manufacturing time and cost. Using a rapid prototype as a communication tool between engineering and manufacturing can lead to simple design changes that yield tremendous savings in the manufacture of the product.
- When rapid prototyping is not a suitable tool for the desired quantity of prototypes, or it does not offer the required material properties, the prototypes are frequently used as patterns for the formation of molds. A common application is the use of rapid prototyping to create a pattern for rubber molding.
- Bridging the product development and manufacturing processes, rapid tooling creates prototypes, short runs of production items, and even full manufacturing runs. Indirectly, as a pattern generator, or by directly creating a tooling insert, the rapid prototyping device crafts the mold for processes like injection molding, investment casting, or reaction injection molding (RIM).

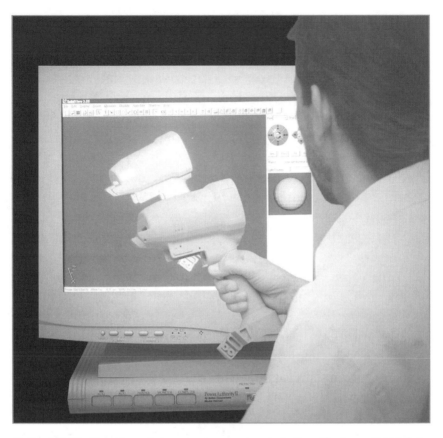

FIGURE 2-12. Rapid prototypes are constructed directly from CAD data, which allows CAD data verification. *(Courtesy Wohlers Associates, Inc.)*

Other Applications

Rapid prototyping is used in the following unique applications.

- Manufacturers would be happy to eliminate the expense and burden of carrying and managing finished goods inventory. Likewise, consumers could be enthralled with custom made, one-of-a-kind products built just for them. These are two goals of rapid manufacturing, the application of rapid prototyping to the production of end-use products. Rapid manufacturing produces finished goods on demand without tooling, molding, or machining.

- Pre-surgical models are one application for rapid proto-typing in the field of biomedicine. A prototype aids surgeons in visualizing, understanding, and appreciating the challenges that they will face with the first incision. Reconstructive surgery, implantation, and even the separation of conjoined twins has benefited from rapid prototyping.

- As those in product development need to visualize a design through a physical representation, so do architects and their clients. Elaborate designs gain physical form so that all involved in their acceptance or modification can reach an agreement (see *Figure 2-13)*.

- Research into the application of rapid prototyping techniques to building construction shows promise. Using the additive process to apply building materials, such as cement on a layer-by-layer basis, facilitates the rapid construction of dwellings with freeform shapes.

- Sculpture and jewelry are two examples of the application of rapid prototyping to artwork. In some cases, the rapid prototype is the final result, while in others the prototype is a molding pattern for the manufacture of the item (see *Figure 2-14*).

- Combined with reverse engineering, rapid prototyping is applied to the science of the study of ancient and modern man. A physical item is scanned in to create an electronic representation. In some cases, the prototype may serve as the skeleton upon which a sculptor constructs a life-like image. In other applications, the rapid prototype allows multiple copies of skeletal remains to be analyzed and displayed.

- Using the techniques of the archeologist, rapid proto-typing has been used to solve murders. From human remains, a scan creates the electronic representation,

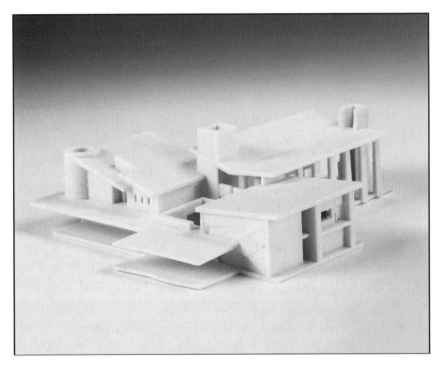

FIGURE 2-13. Architectural models are a growing application for rapid prototyping. *(Courtesy Z Corporation)*

which in turn creates the prototype. Skillfully applying clay to the skeletal structure, a life-like portrayal emerges, aiding in victim identification.

RAPID PROTOTYPING SYSTEMS CLASSIFICATIONS

There are many ways to classify the various rapid prototyping technologies. The devices could be classified by characteristics such as: application, material, cost, speed, size, and many others. To date there is no single, generally agreed upon classification system for rapid prototyping.

FIGURE 2-14. This sculpture from Bathsheba Grossman is a bronze casting produced from a pattern generated with the powder-binder printing process. *(Courtesy Bathsheba Grossman)*

To facilitate discussion throughout this text, a classification system is proposed. Combining various elements and parameters of the commercially available technologies, four classes of technology are defined. These classes consider cost, operating environment, operational overhead, and primary application.

1. Operating in an office environment with little demand for training and oversight, 3D printers function as peripheral devices. From CAD software applications, these devices generate multiple iterations of a design early in the design's life cycle. Combining low purchase price and low operating expense, 3D printers are cost-effective for concept models and early form

and fit analysis. One example of a 3D printer, a Dimension™, is shown in *Figure 2-15*.

2. Enterprise prototyping centers, the original class of rapid prototyping systems, offer a breadth of applica-

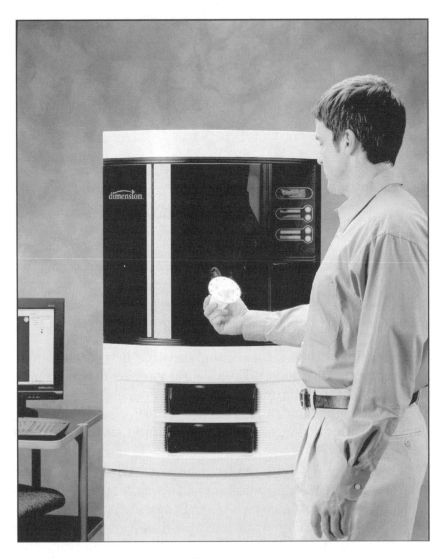

FIGURE 2-15. The Dimension™ from Stratasys is one of a growing number of 3D printers. *(Courtesy Stratasys, Inc.)*

tions and materials. While more expensive, in both purchase and operating expense, these devices offer greater capacity, higher quality, and greater process flexibility. Due to both size and environmental factors, they are best suited for either a lab or shop floor environment, as shown in *Figure 2-16*. Additionally, this class of technology most often demands a dedicated

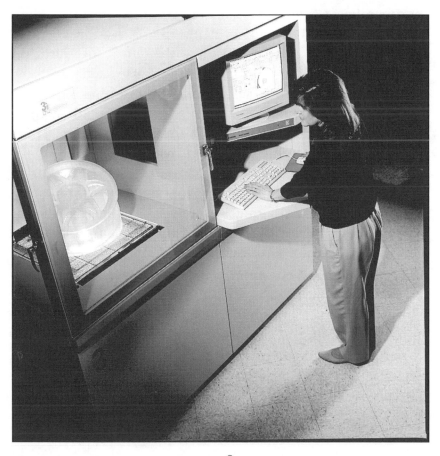

FIGURE 2-16. Systems like the SLA® 7000 from 3D Systems are best suited for a lab or shop-floor environment due to their size and demand for environmental controls. *(Courtesy 3D Systems)*

and well-trained staff to support the needs of all departments throughout a company.

3. Direct digital tooling devices address the rapid tooling applications. Currently, this class is a subset of the enterprise prototyping centers. Few technologies exist that have a single focus on tooling. Most attempt to address both tooling and prototyping requirements. Like the enterprise prototyping center solutions, this class is more expensive, demands a trained staff, and operates in a shop floor or lab setting.

4. Direct digital manufacturing devices address rapid manufacturing applications—the use of additive processes for the production of end-use items. Without intermediate processes, such as tool making, individual components and entire products are created directly from digital data. While some enterprise prototyping center solutions have proven viable in this class, there are no systems that address rapid manufacturing as the primary application. In the future, companies will design and develop these systems with the controls, process, and feedback mechanisms demanded of a production environment machine.

Since this is an early attempt to classify the rapid prototyping technologies, it is wise to further clarify the definition of 3D printing. Some believe that it encompasses the entire scope of rapid prototyping. However, this view is a by-product of the early days of the technology when what became enterprise prototyping centers were described as 3D printing devices.

RAPID TOOLING AND RAPID MANUFACTURING

While the overall discussion of the technology focuses primarily on prototyping, rapid tooling and rapid manufacturing are so intertwined that they cannot be overlooked. In the future these applications may overshadow rapid prototyping.

Rapid Tooling

Soon after the birth of rapid prototyping, the industry realized that addressing only the direct production of a physical prototype was far too limiting. The early applications included silicone rubber molding and investment casting. Each of these long-established processes had a common barrier, the length of time required to fabricate a pattern. In each of these applications, the tooling is crafted by forming a material around a model (pattern) of the desired object. After forming the tool, the pattern is removed or evacuated, leaving a cavity into which plastic or metal is cast. In the early 1990s, the use of rapid prototypes cut the pattern development time for these two processes.

From the early successes, the number of applications for rapid prototyping as a pattern generator has grown. Epoxy tooling, plaster mold casting, sand casting, reaction injection molding, spray metal tooling, and other applications sought to decrease the total lead time by reducing the time required to fabricate a pattern. In some ways, rapid prototyping created renewed interest in many of these processes. In the mid-1990s, the term "rapid tooling" became commonly used. At first, it was applied to these indirect, pattern-based methods. However, the term began to take on new connotations as the goal of rapid tooling shifted.

In the mid-1900s, it was common for a tool for injection molding (and other processes) to take six to eight weeks for even moderately complex designs—much longer for larger or more complex parts. In addition, these were not production tools, but prototype or bridge tools. The new focus of rapid tooling was to slash lead time to allow a design team to secure hundreds of molded parts in just a few weeks. Additionally, the goal was to produce truly functional prototypes in end-use materials. Thus, rapid tooling was produced by two methods, both of which continue to be used today. These are:

1. Indirect methods, where tooling is constructed from a pattern generated by rapid prototyping.

2. Direct methods, where tooling is constructed by the rapid prototyping system.

Just as the approach varies, so does the implicit deliverable of rapid tooling. Some believe that rapid tooling must offer end-use materials. Others consider any mold or tool constructed by rapid prototyping as a rapid tooling system, independent of the prototyping material. For example, the use of silicone rubber molds is common for the production of 5–50 prototypes in thermoset resins. While the production material for these parts is likely to be a thermoplastic resin, many still refer to silicone rubber molding as a rapid tooling solution.

Reality has been unable to meet desires. For the direct methods, rapid tooling proved to have significant limitations, including poor accuracy, surface finish, and tool life. While rapid tooling often slashed delivery time and cost, many could not justify its use in light of the limitations. As a result, rapid tooling became a niche solution, one with little broad appeal.

In the meantime, conventional processes rose to meet the competitive threat. For example, machined aluminum tooling for injection molding no longer takes six to eight weeks. Today, it takes just two to four weeks. While rapid tooling has improved in terms of quality, surface finish, and tool life, the advances have done little to overcome the perceived advantages of machined tooling. Moreover, the gap between cost and delivery time is often imperceptible.

A new direction that holds great promise is for rapid tooling to deliver advantages that are impossible with machined tooling. The two key areas of research are in conformal cooling and gradient materials. Both seek to dramatically decrease cycle times in the molding process. In the case of gradient materials, this is achieved while improving tool life.

Rapid Manufacturing

Today few consider rapid prototyping as a viable option for manufacturing end-use products. Many view it as a future possibility. However, necessity and innovation have already yielded beneficial rapid manufacturing applications.

Few industries or applications are required to meet specifications as stringent as those applied to military aircraft and space vehicles. So, some find it surprising, even amazing, that rapid prototyping has already been used in fighter aircrafts, the space shuttle (see *Figure 2-17*), and the space station. Fully qualified for flight, the rapid manufactured parts have yielded time and cost savings. For the limited number of units in production, tooling and molding were much more expensive and time consuming.

Coming down to earth, rapid prototyping has been applied to other products with extremely low production volumes, such as race cars. Both directly and indirectly, rapid prototyping is used to construct metal and plastic components for NASCAR and Formula 1 race cars. In this fast paced environment where every ounce of weight reduction is critical, race teams have found that rapid proto-

FIGURE 2-17. To reduce time and expense, the space shuttle relies on rapid manufacturing for flight-approved components. *(Courtesy National Aeronautics and Space Administration [NASA])*

typing allows them to quickly realize production parts that improve performance.

Obviously, aforementioned representative examples are unique. Each has production runs measured in tens, not tens of thousands, and each faces design challenges not common in the typical consumer or industrial product. Yet, these everyday applications can also benefit from rapid manufacturing. Innovative applications are arising every day as companies consider the advantages and possibilities rather than the obstacles and risks. As more companies explore the opportunities, and as the technology develops into a suitable manufacturing process, rapid manufacturing will grow beyond a niche application to become a routinely used solution.

FUTURE OF RAPID PROTOTYPING

Rapid prototyping is a relatively new technology that demonstrates many of the characteristics of a technology that is early in its life cycle. Most notably, no one can predict with great accuracy what the future holds. How will the technology evolve in the future? What technologies will be around? What will it cost? These are just a few of the questions that can be answered only in due time. Yet, there are some speculations that can be offered with a degree of confidence. The first is that change will be the only constant. The second is that rapid prototyping, in some form, is here to stay.

As with many new technologies, research and development is at a significant level. New methods, new applications, and new materials are in labs around the world. Many more will follow. What this means to the users of rapid prototyping is that the future is likely to reveal not only many small, incremental changes, but also a handful of disruptive technologies that change the game entirely. Be it five, ten, or twenty years into the future, rapid prototyping will have broader application, wider acceptance, and greater impact on industry.

Although it appears contradictory, at the same time that R&D increases, the rapid prototyping industry will move toward standardization. As seen in the brief description of the four leading technologies, there is little commonality among these processes. As we progress into the future, the consumer will select the standards by voting with dollars. Over time, the consumers will clearly indicate what is most beneficial and desirable in terms of rapid prototyping. Without standards, no technology can flourish. The risk of obsolescence and the fear of choosing poorly delay entrance into a new technology. The lack of standards also bars further decrease in price with added performance. In the coming years, standards for the process and deliverables will arise. These standards will also yield clear classification of rapid prototyping technologies.

It is unlikely that any individual can go through a single day without benefiting from rapid prototyping technology. Thousands of products have been improved with rapid prototyping. It has allowed companies to respond to competitive threats by decreasing the time it takes to deliver a product via conventional processes. This trend will continue and grow.

REFERENCES

Hull, C. 1986. *Apparatus for Production of Three-Dimensional Objects by Stereolithography.* U.S. patent 4575330, Issue date 3/11/86. Washington, DC: U.S. Patent and Trademark Office.

Jacobs, P. F. 1992. *Rapid Prototyping & Manufacturing; Fundamentals of Stereolithography.* Dearborn, MI: Society of Manufacturing Engineers.

Wohlers, T. 2003a. www.wohlersassociates.com. Fort Collins, CO: Wohlers Associates, Inc.

Wohlers, T. 2003b. *Wohlers Report 2003*: *Rapid Prototyping & Tooling State of the Industry Annual Worldwide Progress Report.* Fort Collins, CO: Wohlers Associates, Inc.

BIBLIOGRAPHY

Global Alliance of Rapid Prototyping Associations (GARPA), www.garpa.org. Eight Mile Plains, Australia: QMI Solutions.

CHAPTER 3

The Rapid Prototyping Process

To understand the origins of the strengths and limitations of rapid prototyping requires an understanding of the process. This information also serves as a foundation for technology evaluations and as an overview of the operational demands.

The uninitiated may perceive rapid prototyping as a push-button technology that offers instantaneous output from a computer-aided design (CAD) system. For those who own and operate rapid prototyping systems, there is an awareness that much more goes into the process. While some technologies approach push-button simplicity and some prototypes can be produced in only one hour, much more time, effort, and consideration are required for the majority of parts.

The description of the process that follows is thorough but not complete. From a hands-on, user perspective, the rapid prototyping process is described in detail. It offers insight into the details of the operation that only experienced users would know. Yet, with the multitude of processes, the information is not a detailed account of each technology. Rather, it is a generalization of the process for all systems.

OVERVIEW

Although each rapid prototyping technology has its own unique methodology, there is commonality in the general process and workflow for producing rapid prototypes. This process consists of five primary steps, which are:

1. STL file generation
2. File verification and repair
3. Build file creation
4. Part construction
5. Part cleaning and finishing

With the exception of the last step, cleaning and finishing, the process is highly automated and requires little time on the part of the system operator or technician. In fact, it is quite common that one individual performs all of this work—with the exception once again of cleaning and finishing—for multiple rapid prototyping machines.

In the early days, rapid prototyping operations demanded highly skilled operators because building a good prototype was as much art as it was science. However, rapid prototyping is evolving from devices that require highly trained personnel to those that many in the organization can operate. As industry standards are developed and system advancements further address user operations, there is the likelihood that in the future many systems will be as easy to run as a copy machine. However, few systems currently offer the simplicity of a copy machine. Even though operations and processes have improved, most of today's systems continue to require the skill set of a technically oriented operator with problem-solving abilities.

When evaluating the five primary steps at the detail level, it may appear that the process is laborious and time consuming. However, the reality is that the entire operation can be completed efficiently and quickly. For the typical prototype, the five steps would take the following amount of time:

1. Generate STL file: 1–10 minutes
2. File verification and repair: 5–30 minutes
3. Build file creation: 15 minutes–1 hour
4. Part construction: 30 minutes–48+ hours
5. Part finishing: 15 minutes–4 hours

With rapid prototyping, it is feasible to complete a prototype in little more than one hour. Yet, some prototypes may take days to complete and deliver. This large variance is the product of the many variables that contribute to the time required for prototype production. The key variables that determine the time are:

- size of the prototype (both height and volume)
- system used
- material used
- level of finish

A part that takes one hour to complete will be small. For some technologies, "small" may imply something the size of a button. For others, a small part could be the size of a cell phone housing. In addition, the one-hour prototype would dictate that the part be delivered with some cleaning but no finishing.

PROCESS DETAIL

A detailed description of the rapid prototyping process offers insight into many aspects of the technology. The discussion offers an understanding of the origin of rapid prototyping's benefits, strengths, and limitations. It also allows an appreciation of the operational considerations for successful and efficient rapid prototyping builds.

STL File Generation

CAD

Rapid prototyping requires unambiguous, three-dimensional digital data as its input. This dictates creation of a 3D computer-aided design (CAD) of the prototype prior to starting the rapid prototyping process. *Figures 3-1* and *3-2* show shaded images of 3D solid models for a foam-dispensing unit that was prototyped with fused deposition modeling.

For those who rely on 2D CAD for product design and definition, there will be additional expense and delay in the prototyping process. Prior to beginning prototype construction, the 2D data must be converted to a 3D file. Since rapid

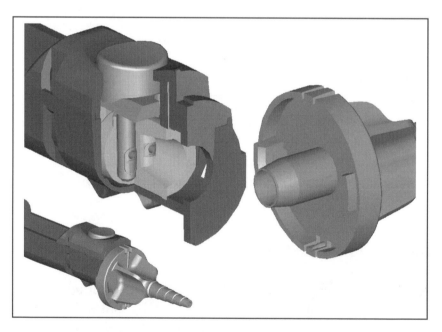

FIGURE 3-1. This 3D CAD solid model of a dispensing nozzle for a two-part foam-dispensing unit (showing both the full nozzle and a cut-away view) was used to create STL files for the fused deposition modeling process. *(Courtesy Leyshon Miller Industries)*

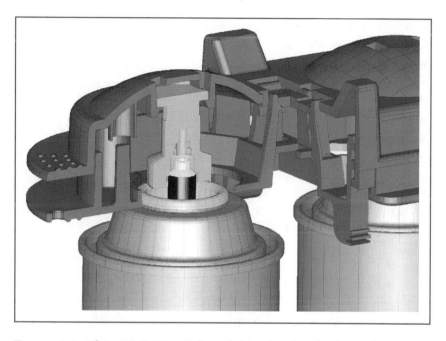

FIGURE 3-2. This 3D CAD solid model is also for the foam-dispensing unit. Two of the components were prototyped with fused deposition modeling, while the valve was CNC machined. *(Courtesy Leyshon Miller Industries)*

prototyping is most powerful for complex designs and CAD programming time is primarily a function of the design's complexity, the data conversion process can take longer and cost more than the prototyping effort. As a result, many cannot justify the use of rapid prototyping when the starting point is a 2D CAD file.

For 3D CAD, there are three modeling options: wireframe, surfaced wireframe, and solids. The first option, wireframe modeling, is not suitable for rapid prototyping applications. Constructed with only edge definition, the wireframe model does not offer the full definition required for prototype development. Both the surfaced wireframe model and solid model are applicable to rapid prototyping, and the selection of the method is purely a personal choice.

However, many find that building a good, clean, usable model with surfaced wireframe can be challenging, difficult, and painstaking.

The quality of the CAD model is critical to rapid prototyping. Modeling techniques and design short cuts that can be accommodated by other manufacturing processes may not be appropriate for rapid prototyping. To generate a usable file for rapid prototyping, the CAD data must be "watertight"—no gaps, holes, or voids—and the geometry must not overlap. Any defect in the model, even those that are not visible to the eye, may produce poor results in the rapid prototype or make the file unusable.

Since the launch of the rapid prototyping industry, all commercial 3D CAD systems for business applications have developed software that allows the export of a stereolithography (STL) file, the neutral file format used by all rapid prototyping systems. This generic functionality means that the selection of the CAD tool is driven not by rapid prototyping requirements but rather by features and functionality for the design task. This was not true in the early days. A decade ago, some CAD systems may not have had STL capabilities, may have generated poor STL files, or may have charged as much as $5,000 per license for STL output functionality. In response to this early technology gap, software developers created applications that produce STL files from surfaced international graphics exchange specification (IGES) data. IGES files, like drawing interchange format (DXF) files, are neutral, allowing file passing between dissimilar CAD systems. These IGES to STL conversion tools are still available today, and while not as critical as in the past, many rapid prototyping operations continue to use them.

STL File Creation

The STL file is a neutral file format designed such that any CAD system can feed data to the rapid prototyping process. All rapid prototyping systems accept this file (see *Figure 3-3*). Its use has also spread to other applications, including machining and electronic collaboration.

STL Defined

Even though the STL file is widely used, few agree on the definition of the acronym STL. Suggestions include standard triangle language, stereolithography language, and stereolithography tessellation language. Chuck Hull, the inventor of stereolithography and 3D Systems' founder, reports that the file extension and acronym stand for stereolithography.

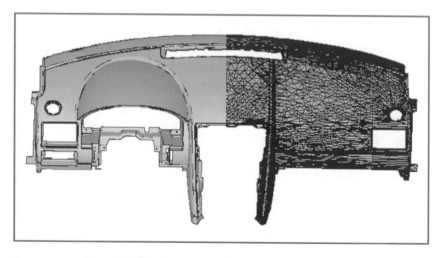

FIGURE 3-3. The STL file is a mesh of triangular elements depicted here on the right half of the digital model. *(Courtesy Materialise)*

The STL file bridges the communication barrier between all CAD systems and rapid prototyping devices. As anyone who has attempted to transfer CAD data can attest to, dissimilar CAD systems are unlikely to offer direct transfer of data. In previous years, the CAD-to-CAD data transfer issue was addressed with two neutral file formats that became industry standards: IGES and DXF. In the mid-1990s, another cross-platform file architecture was announced. In December 1994, the International Organization for Standardization (ISO) adopted the standard for the exchange of product model data (STEP).

The STL files, much like IGES, DXF and STEP, allow all CAD systems to feed data to rapid prototyping equipment. However, the STL file format is much simpler than that of the other neutral file formats since it contains much less information and eliminates the need for a translator on the receiving end. The trade-off for this simplicity is that the STL file lacks the data architecture for advanced modification and manipulation. Anything beyond scaling and sectioning of the model may require a return to the original CAD file.

The STL file approximates the geometry in the CAD file. Much like a finite element model, a mesh of elements represents the part, as shown in *Figure 3-4*. However, the similarities end there. The STL file uses a simple triangular mesh to approximate the bounding surface of the part. Additionally, the STL file strives to accurately approximate the data with as few elements as possible.

Exported in either binary or ASCII format, an STL file is simply a listing of the coordinates of each vertex of the triangles in the mesh. Combined with a surface normal, a vector indicating the outward direction, the listing of all of the triangular elements provides a complete description of the 3D CAD data to be constructed.

When exporting an STL file, the goal is to balance model quality and file size. This is done by dictating the allowable

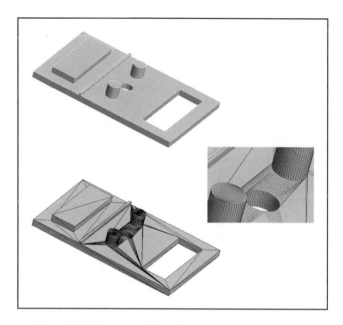

FIGURE 3-4. The 3D CAD model (top) is exported to generate an STL file (bottom). The triangular elements of the STL mesh are shown. A high concentration of triangles is required to accurately depict cylindrical features in the blown-up view (right).

deviation between the model's surface and the face of the triangle. Although there are various terms for this deviation—chord height, facet deviation, and others—each CAD system allows the user to identify the allowable gap between the triangle face and the part surface. *Figure 3-5* depicts an STL representation of a cylinder where the facets would be clearly evident in a rapid prototype. With smaller deviation, model accuracy improves. In general, a facet deviation of 0.001–0.002 in. (0.025–0.05 mm) is sufficient for the rapid prototyping process. Larger deviations may result in visible facets on the prototype that detract from aesthetic appeal and accuracy.

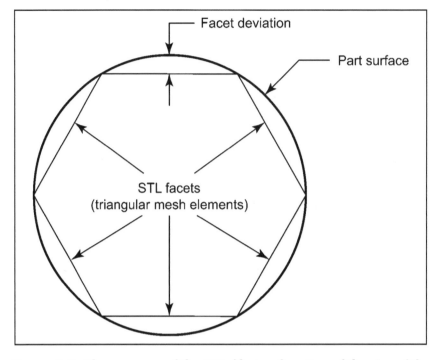

FIGURE 3-5. The accuracy of the STL file is a function of the size of the facets and the resulting deviation from the part surface.

File Verification and Repair

The original CAD model and the STL generator can yield defects in the STL file that will affect the quality of the prototype or prevent its use. Common defects include near-flat triangles, non-coincident triangles, overlapping triangles, inversed surface normals, and large, visible facets.

Problems occur even when best practices have been observed in the creation of the CAD data and STL file. In day-to-day rapid prototyping operations, a significant percentage of STL files will be corrupt. To fix these file problems, detection and repair software is available from rapid prototyping system manufacturers and third parties.

The two basic rules for a good STL file are: 1) adjacent triangles must share two vertices; 2) surface normals must point away from the volume of the part. When these rules are broken, the file is corrupt and unusable by a rapid prototyping device.

Prior to processing STL files for building, a software verification program analyzes the data. These programs report file defects and their severity, allowing the operator to determine if the data is usable. When a bad file is discovered, it is processed with an STL repair program. Such applications attempt to resolve file deficiencies with the correction of surface normals by filling gaps between triangles and elimination of near-flat triangles. In most cases, the corruption can be repaired within the STL file. However, in some instances, repairs must be made to the original CAD file and a new STL file exported. The necessity to return to the original CAD file is often the result of poor CAD modeling techniques. If the facets of the STL file are too large, making them visible in the prototype and degrading accuracy, the process must return to the original CAD file for the generation of a new STL file.

Build File Creation

To prepare the STL files for building, several steps are required. While this process may appear to be time consuming, it typically takes as little as a few minutes and no more than an hour. The steps in file processing include part orientation, support structure generation, part placement, slicing, and build file creation.

Part Orientation

Careful consideration of the orientation of the prototype is important to balancing part quality and machine time. In most rapid prototyping systems, the height of the part has a significant impact on build time. As the height increases, build times become longer. However, this must be balanced with part quality since accuracy, surface finish, and feature definition will vary with respect to the plane in which a feature is located.

The Z axis (the vertical axis in a rapid prototyping system) of any rapid prototype is an important consideration when preparing a build. The combination of part height and layer thickness is a predominant factor in the time it will take to construct the prototype. As the height increases or the layer thickness decreases, the number of layers in the rapid prototype increases. An increase in the number of layers extends the build time. This contributes to operational overhead. Each layer of a prototype, for most systems, contributes some amount of unproductive time. In effect, there are small setups between each layer. These may include: platen repositioning, calibration, and material leveling. Even if these steps are completed quickly, they can add an appreciable amount to the build time. For example, a 5-in. (127-mm) tall part, with 0.004-in. (0.10-mm) layers, and 15 seconds of setup time per layer, would have 5.2 hours in operational overhead.

Since all rapid prototyping systems construct models layer-by-layer, all prototypes will exhibit some degree of "stair stepping" (see *Figure 3-6*) unless all features are parallel with the horizontal or vertical planes. As the layer

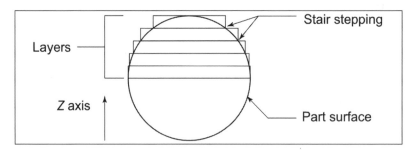

FIGURE 3-6. Stair stepping is the result of an approximation of geometric features by the horizontal layers (slices).

thickness decreases, the stair-stepping effect is minimized, surfaces become smoother, and features become more accurate. Stair stepping is analogous to aliasing in a 2D representation of a line that is neither vertical nor horizontal. In the 2D world, such a line is represented by discrete moves in the X and Y axes, creating a stepped representation of what should be a smooth, angled line. In rapid prototyping, any feature that angles or curves away from the Z axis is interpreted as a discrete shift, not a smooth transition, between layers.

If the prototype is intended for an application that requires smooth surfaces, for example a pattern for tooling, increased stair stepping will increase the amount of time in part finishing. Additionally, an increase in the amount of finishing can decrease the accuracy of the part. With heavier sanding, more material is removed to yield a smooth surface.

Orientation affects quality for some features aligned in the horizontal plane. For example, a through hole oriented with its centerline along the X axis will be approximated with the stepped layers. Not only will the hole have stepped surfaces, but it can take on an oval shape, as shown in *Figure 3-7*. The oval shape occurs for two reasons. First, the uppermost and lowermost points of the hole will be flat, represented by the top or bottom of a layer. Second, layer interpretation may be different for the upper and lower halves of the hole. The lower half may have stair stepping where the bottom of each layer is coincident with the diameter, and the upper half uses the top of each layer.

Beyond the aforementioned aspects of orientation, each rapid prototyping system will have its own unique considerations. For some systems, the top-most and bottom-most surfaces are the smoothest and most accurate. For others the converse is true. As will be discussed, orientation will also prescribe the amount and location of support structures.

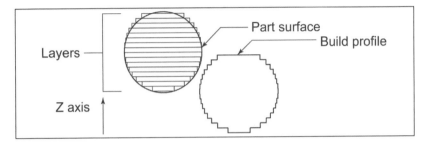

FIGURE 3-7. The effect of stair stepping can create an oval shape for some geometry.

Balancing Time and Quality

A simple example of the balance between time and quality is illustrated with the plastic body of a ballpoint pen. For best quality, the pen should be oriented vertically. This allows both the outer diameter (OD) and inner diameter (ID) to be accurately represented with no stair stepping or oval shape resulting. Yet, in a stereolithography system, for example, a vertical orientation would yield a build time of five to ten hours, depending on the system, layer thickness, and material. Laid on its side, the build time would be reduced to one or two hours. Yet, the pen body will be stair stepped and ovalled on both the OD and ID.

Since overhead is generally a function of the overall build height, and not the number of parts or the cumulative height, part quality and build time can be balanced with the use of a full platen of parts. For example, when constructing a single pen in a vertical orientation, the build time could be five hours. If two pens are constructed at the same time, the cumulative build time would be in the neighborhood of 5.5 hours. If 100 pens were built, the total time would be 50 hours, or just 30 minutes per pen, which is nearly identical to the time for one pen constructed in a horizontal orientation.

Support Structures

Rapid prototypes can start as a powder, liquid, or semi-molten material that is not constrained in a tool or mold.

While some systems, like selective laser sintering and powder-binder printing surround the rapid prototype within the unused, excess material, most systems construct the prototype within a liquid environment or in open space. (Z Corporation refers to their systems as a 3D printer, not by a process name. Since the term "3D printer" is generically applied to many technologies, this text refers to the technology as powder-binder printing to reflect the modified inkjet printing process used in the company's technology.) To prevent the prototype from shifting during the build and to eliminate sagging or slumping of features, rapid prototyping systems rely on support structures. An example of a support structure is shown in *Figure 3-8.*

Supports serve two functions: rigidly attaching the part to the build platen and supporting any overhanging geometry. To fixture the prototype, support structures are added to the base of the prototype. During part building, the base supports are constructed first. They firmly attach to the build platen to provide a pedestal upon which the prototype is constructed. These base supports also provide a gap between the part surface and the platen. This allows removal of the prototype from the platen, often through aggressive mechanical means, without damage to the part surface. Without these base support structures, a prototype in stereolithography, for example, would shift within the build area as the liquid resin is disturbed by platen displacement.

The other purpose of support structures is to retain a feature that has no previously solidified material below it. Examples include protruding bosses, large circular holes, and horizontal flanges. By adding support structures below these overhanging features, a scaffold is constructed prior to building the feature, and it is on this scaffold that the feature rests. This type of support structure extends down to the platen or to any solid geometry located between the supported feature and the platen.

FIGURE 3-8. The support structure for a stereolithography part can be seen between the bottom of the wheel and the building platform. *(Courtesy 3D Systems)*

Supports take on a multitude of designs, each predicated by the system manufacturer. Within any given technology, there are a number of configurations to accommodate various build scenarios. The most common and widely used design is the egg crate or checkerboard pattern. In this configuration, the supports are thin, widely spaced vertical walls. The walls are oriented in both the X and Y axes to create a grid, or checkerboard. In this design, each support wall has a thickness of just a few thousandths of an inch— typically 0.010–0.020 in. (0.25–0.50 mm). Other support structure configurations include solid masses of material, alternating patterns of solid and hatched material, or small columns of material.

Self-supporting prototypes use an entirely different strategy. These technologies take advantage of excess, unused material to contain and support the prototype. In selective laser sintering, for example, each new layer starts with the addition of a fresh layer of powder. After sintering a layer of the prototype, the unsintered material remains. Surrounding the prototype with the unsintered material provides the necessary support. When a build is completed, the prototype is contained in a "cake" of powder that is removed in the cleaning stage. The same is true for powder-binder printing, as shown in *Figure 3-9*.

Not all overhanging features require support structures, for example a canted wall or a through hole. In the instance of a canted wall, the equipment vendor will have a predetermined maximum angle to which the wall can be built without supports. For smaller angles, the previously solidified material will support the next layer. This is also true for through holes with internal diameters that do not exceed system specifications. For smaller holes, the previously solidified material supports subsequent layers.

For many prototypes, the details of support structure configuration can be ignored. These rules and guidelines are contained within automated support generation programs. After orienting the prototype, the program will create the required support structure in only a few minutes. While the basic automated approach to support structures is quick and effortless, advanced users may take a different approach that requires thought, time, and manual intervention. Advanced application of supports addresses the impact of supports on time, cost, and quality. The goal of automated support generation programs is to ensure that a part builds successfully. But this conservative approach can increase machine time, material cost, and part finishing time. For all systems, support structures consume additional material, and for most systems, support structures increase the time to construct the prototype. While this

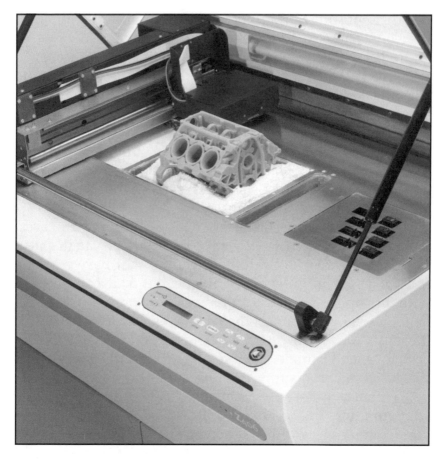

FIGURE 3-9. The engine block rests on the bed of powder that supported the part during the construction process. *(Courtesy Z Corporation)*

additional time and cost may be insignificant for small parts, it can add hours and hundreds of dollars to a large prototype that requires a substantial number of support structures. For these large prototypes, minimization of the number of support structures can be very beneficial.

The other area of consideration is the support structure impact on finishing time, surface quality, and feature retention. When a build is complete, the rapid prototype is cleaned and finished, which requires the manual (in most cases) removal of the support structure. With an increase in the number of supports, especially surrounding small features, the time to remove them will increase. Also, when stripped from a part's surface, the support structures often leave a visible mar or remnant. These must be eliminated for visual appeal and accuracy through manual finishing efforts with sandpaper and files.

The last support structure consideration is accessibility. If a bench technician cannot gain access to the support structure, it will remain within the part. A good example of this scenario is a valve body. If the intake and exhaust ports are not large enough to allow access to the central chamber, the supports may have to remain in the valve. In some cases, the supports may be picked out with files and tweezers, but the internal surface will not be smooth. For flow testing of the valve body, neither scenario is acceptable. In these circumstances, support structures may force the use of another rapid prototyping technology.

When considering part orientation, support structures should be a component of the decision-making process. As previously stated, support structures that surround and contact small features make the removal process time consuming, and they can also affect quality. When support structures surround or contact small features, removal can lead to damage. It is common for thin ribs or small posts to be broken off during the removal process. Thus, a part may be oriented such that supports around small features are minimized or eliminated.

A cell phone housing is a good example of a support-structure-based part orientation decision. The exterior of the housing has easily accessible surfaces and a minimal amount of fine detail. The interior of the housing is a different story. It contains small support ribs, mounting bosses, and stand-offs. (The exterior is the upward facing surface in the build.) The interior of the part requires supports around and contacting these small features. Upon support removal, they may be damaged. However, there is a tradeoff if the orientation is inverted and the exterior is supported. The visible mars and remnants of the supports will degrade the visual appeal and surface quality. The selection of the supported surface will, as a result, depend on the intended application of the prototype and operational preferences.

Part Placement

Rapid prototyping, unlike many prototyping methodologies, can construct multiple parts concurrently. The mix and number of parts has only a few limitations: All parts must reside within the bounding box of the usable build envelope and cumulative file size must not exceed system maximums. For some technologies, there is one additional limitation: Parts should not be stacked on top of each other. If these constraints are satisfied, any combination of prototypes can be constructed in one machine run. This could be dozens of the same part, a collection of different parts, or even multiple parts from multiple projects. To capitalize on this unique situation, there should be careful consideration of part placement, part inclusion, and the overall build layout. These decisions can impact operational efficiency, throughput, build times, and part quality.

Using the vendor-supplied software application, part files that have been oriented and supported are placed within a bounding box that represents the build envelope. Each part file is given its own area of the build footprint, ensuring that it does not impinge on the footprint of any other parts. With these basic rules accounted for, a build can be prepared. However, advanced system operation requires additional consideration of part accuracy and machine efficiency. For some rapid prototyping technologies, like stereolithography and selective laser sintering, a part's position within the build envelope can affect dimensional accuracy and feature definition.

In stereolithography, the laser used to cure the resin originates from a fixed position. Scanning mirrors redirect the laser for travel in the X-Y plane. When curing material in the very center of the vat, the laser spot is a perfect circle. However, as the beam traverses toward the edges of the build area, it takes on an elliptical shape. While the deviation from circle to ellipse may be small, it can add material to the part, making it larger than the design specification.

In selective laser sintering, the same ovaling occurs. However, it is overshadowed by temperature variation within the powder bed. Parts located at the outward extent are the most likely to demonstrate curling and warping of the part's surface. As the part construction occurs, a variation in powder temperature builds. The imparted energy of the CO_2 laser and the heat retention characteristics of the powder create a higher temperature in the center of the build piston. The powder temperature decreases as it nears the edges of the build area. This gradient may cause geometry located on the outer edges of the build area to distort.

The quantity and variety of parts to include in a build directly impacts build time, throughput, and operational efficiency. Since build time comprises part generation time and operational overhead on each layer, it is best to run these systems with full utilization of the available build envelope. By tightly packing parts onto the platen, as shown in *Figure 3-10*, the overhead component of time is amortized over all parts. Overhead is not cumulative, as illustrated in the ballpoint pen example. Packing the platen yields a build time that is less than the cumulative time of individual parts run separately.

To take full advantage of space in the build envelope, the mix of parts should have relatively the same Z height. If, for example, a build is comprised of one part that is 10-in. (254-mm) tall, and all others are less than 2-in. (51-mm), the advantages of packing the platform are not nearly as great as if all parts were in either the 2-in. (51-mm) or 10-in. (254-

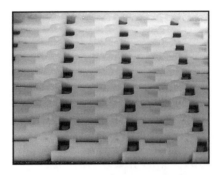

FIGURE 3-10. Efficient use of the entire build area reduces total time and cost for many rapid prototyping systems as shown with these fused deposition modeling prototypes. *(Courtesy Stratasys, Inc.)*

mm) range. This is because the 10-in. (254-mm) part carries the full overhead burden for the top 8 in. (203 mm), and shares the burden with the other parts for only the first 2 in. (51 mm), as shown in *Figure 3-11*.

While not always practical, it is best to build parts that have similar Z heights in the same machine run. This offers the greatest operational efficiency and lowest operating expense. However, in the real world, the dynamics and demands for prototypes often dictate operating procedures beyond those of the highest efficiency.

When placing prototypes for a build, delivery requirements must be considered for each part. While it may be most convenient for the machine operator to run one build with the 10-in. (254-mm) and 2-in. (51-mm) part combination, the total lead time will be greater for the small parts. If built without the 10-in. (254-mm) part, the smaller ones could be available for finishing hours sooner.

Slicing

After completing a build layout, the STL files for both the part and supports are sliced into thin, horizontal cross sections. Each cross section represents one layer of the

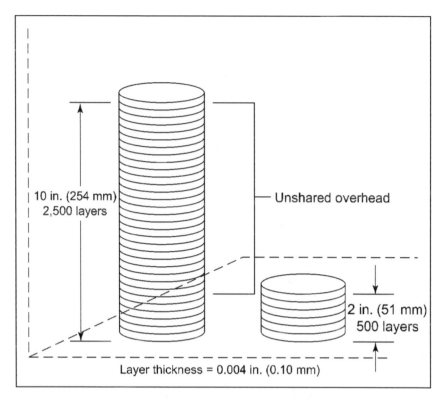

FIGURE 3-11. When prototypes of varying heights are combined in one run of a rapid prototyping machine, some of the operational overhead is not shared across all parts.

building process. In effect, these layers become the "tool paths" that drive the laser, print head, or extrusion tip of the rapid prototyping device.

Commercially available systems offer layer thicknesses of 0.0005–0.020 in. (0.013–0.51 mm), and most offer the user a range from which to select. If the option is available, there are a few considerations when specifying the desired layer thickness.

As discussed previously, layer thickness has an impact on both part quality and build time. Thinner layers yield improved surface quality and dimensional accuracy. They can also reduce the amount of time invested in prototype benching. However, thinner layers increase build time. With an increased number of layers, there is an increase in the time lost to operational overhead. Therefore, when selecting the desired layer thickness, it is important to consider not only the application of the prototype (and the resulting quality parameters), but also the time it will take to construct, and that time's impact on schedules, deliveries, and operational efficiencies.

To maximize quality and minimize build time, it is often possible to specify multiple layer thicknesses for a given part. For example, if a prototype has an area of fine detail or flowing surfaces but the bulk of the part lacks similar detail, a thinner layer thickness may be specified for only the area with detail. The balance of the part would capitalize on thicker layers and reduced time. However, this strategy has one limitation: Variable layer thickness would apply to all parts within the build. At a given Z height, there cannot be two different layer thicknesses.

For the majority of rapid prototyping systems, the common layer thicknesses employed by users are 0.004–0.010 in. (0.10–0.25 mm). This range often yields the best balance between part quality and build time.

The Final Step

The final step in file processing is creating the build files that drive the rapid prototyping system. As with support generation, this can be a simple process when using system defaults to automatically create the data, or it can be a thoughtful process that requires experience and skill. For experienced operators, the standard parameters are modified to decrease build time and improve part quality.

When defaults are used, the build parameters are dictated by the given rapid prototyping system and the material that will be processed. Each class of system within a given rapid prototyping technology will have its own unique build parameters. Likewise, each material has its own specifications.

Considering all technologies and individual systems, the variables are too extensive and diverse to list and discuss. However, in general, these parameters relate to speeds, dwells, path widths, path depths, path overlap, and fill types. Combined, these parameter sets are commonly known as a build style.

Part Construction

Perhaps one of the biggest advantages of rapid prototyping is that the operation is unattended. With few exceptions, rapid prototyping systems can operate 24 hours a day without human labor. The only labor required of all rapid prototyping systems is for machine preparation, build launch, and the removal of the prototypes upon completion.

To prepare for the build, material is added. Some technologies require that the reservoir of material be brought to operating levels, while others only dictate that enough material be available to complete the build. Additional preparation may include a preheat cycle to bring the system up to operating temperature.

Rapid prototype part construction is the inverse to machining in two ways. First, material is added to create the part, not removed. Second, rapid prototyping systems start construction at the bottom of the part. Rather than machining the uppermost surface of a block of material first, rapid prototyping creates the lowermost layer of material and then stacks additional layers on top.

To complete a build, the last layer—the top of the part—is solidified. For many rapid prototyping systems, such as the powder-based systems, the partially constructed prototype is not visible during the build. It is not until the part is completed and the platform raised, that the part can be viewed. What is visible in every system is the uppermost working surface, as shown in the time-lapse image of *Figure 3-12*. Here the build process can be observed. As the laser moves across the parts, the print head deposits binder or the nozzle extrudes semi-molten material onto the top of the partially constructed prototype.

As the rapid prototyping process begins, the lowermost layer becomes the tool path for the laser, print head, or extrusion tip. Each layer is simply a two-axis operation. The slice information defines the boundaries of the profile and the internal fill area. The thickness of each layer is controlled by the laser draw speed, the size of the droplets, or the thickness of the extrusion. For most technologies, the

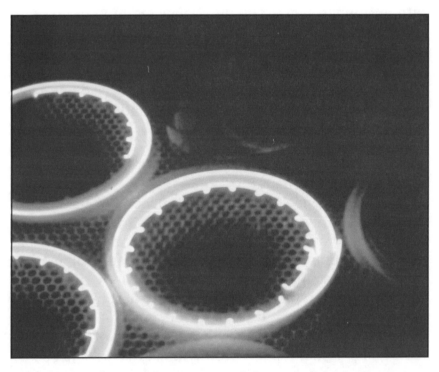

FIGURE 3-12. This time-lapse image of the stereolithography process shows the traces of the ultraviolet laser on the surface of the liquid photopolymer. *(Courtesy 3D Systems)*

boundaries are solidified first and then the internal area is created. The most obvious exception is with material jetting systems; for these, the drop-on-demand technology constructs the boundaries and the fill in a single pass.

The method of solidifying the internal area varies by system and selected build style. For example, stereolithography hatches the fill area with overlapping passes, and these passes are made in both the X and Y axes. Selective laser sintering, on the other hand, uses only a single, non-overlapping pass in one axis.

The length of time to construct prototypes varies dramatically by system, operating parameters, build height, and build volume. While a large, thick-walled part could take three days or more, most machine runs will range from 30 minutes to 48 hours. In addition to the sensitivity to build height (Z axis) and layer thickness, build time is a function of the prototype's volume and the method used to create it. For the curing, sintering, and extrusion processes, the volume of a part contributes significantly to the build time. In jetting technologies, however, this is of little consequence.

For the non-jetting processes, the amount of material between the boundaries defines the number of passes required of the laser or extrusion nozzle. Therefore, the greater the volume of the part, the more time to construct it. Jetting technologies are not nearly as sensitive to the volume. A single pass of the print head deposits a full layer of material. However, the key element of time for all systems is the part's overall size.

Unlike machining, complexity plays no part in time; rather, the size and volume of the part define the construction time. It is for this very reason that the most beneficial applications of rapid prototyping, when compared to machining, are those where size is limited and complexity is great. In rapid prototyping, a part with a volume of 1 in.^{3} (16.4 cm^3) measuring $2\times2\times3$ in. ($51\times51\times76$ mm) takes nearly the same amount of time whether it is a simple box shape or a complex shape with free-form surfaces, undercuts, ribs, and bosses.

After the completion of a layer, the system prepares for the next. The build platen drops down by one layer thickness, leaving a gap between the finished layer and the working surface. For stereolithography, selective laser sintering, and powder-binder printing, raw material fills this gap. For fused deposition modeling, this is an air gap.

With the platen in position, a number of operations are completed before drawing the subsequent layer. These differ greatly by system and may include:

- addition of a new layer of material
- leveling of the material surface
- purging of material from the tool, or
- sensor checks for power, temperature, and other system parameters

This combination of procedures between each layer is what creates the operational overhead component of the overall build time. These procedures are not dependent on the number of parts on each layer, and this is what allows amortization of system time over all parts.

While the lost time between layers may be less than one second, it can approach a full minute. When accumulated over thousands of layers for a build—a 5-in. (127-mm) part with a slice thickness of 0.004 in. (0.10 mm) has 1,250 layers—the overhead time can become appreciable.

With the system prepared for the next layer, the building process is repeated. This subsequent layer bonds to the preceding layer. After the second layer is completed, the entire process repeats for all subsequent layers until the rapid prototype is complete.

Some systems require a cool-down period after completion of the build. This allows the temperature to decrease until the prototypes can be removed without damage or deformation. If an attempt to remove the parts is made too soon, the part may be pliable or not fully cured.

Part Cleaning and Finishing

Cleaning

With the prototype build completed, the manual, labor-intensive processes begin. Contrary to the depiction of rapid-prototyping-like devices in movies and television, the prototypes are not usable directly from the machine. Each prototype must be cleaned, and usually benched, to make it suitable for application and use.

The first step is cleaning. In general, rapid prototyping requires the removal of any excess material and support structures. All other aspects are process dependent. Cleaning methods may include post-curing, chemical stripping, bead blasting, or water jetting. For the systems reviewed in this book, cleaning involves the following steps.

- Stereolithography requires chemical stripping of uncured resin from the part's surface, stripping of the support structure from the part's surface, and curing in an ultraviolet (UV) oven to bring the part to a full cure.
- Selective laser sintering requires removing the part from the powder cake, brushing excess powder from the part's surface, and evacuating powder from within any cavities.
- Fused deposition modeling requires the removal of support structures.
- Powder-binder printing requires brushing excess powder from the part's surface and evacuating powder from within any cavities.

As interest grows in the application of rapid prototyping in a desktop or office environment, the cleaning operations are one of the limiting factors. The process can be messy and may require equipment and chemicals that are suited only for a shop environment.

For some technologies, there may be an additional step required prior to part benching: infiltration. To increase strength and eliminate porosity, some systems, like selective laser sintering and powder-binder printing, benefit from an infiltrant that wicks into the pores of the part and seals the outer surface. There are a number of compounds used as infiltrants, including cyanoacrylate and epoxy.

Finishing

Depending on the application of the prototype, additional part preparation and finishing may be required. Beyond support removal, consideration of the application of the prototype is an important element in deciding the level of finish. Inherently, each prototype will have evidence of stair stepping, and some may have an inherent surface roughness from binding or sintering powders or extruding semi-molten plastic. The question that must be asked is whether this is acceptable. For a simple proof-of-design concept model, these deficiencies are likely to be acceptable. For a pattern or painted showpiece, they must be eliminated.

Commonly, benching of the prototype includes sanding. For exceptional quality, sanding may be accompanied by the filling of mars and pocks. These operations require the hand tools common to any model shop. Sandpaper, files, and surface profilers are just a few of the tools that may be used. The list may also include paint booths, bead blasters, and Dremel® tools, among others.

There is no doubt that part finishing is a labor-intensive process that requires a unique skill set (see *Figure 3-13*).

FIGURE 3-13. When the rapid prototyping process is complete, parts are often finished to specification, a process that is not automated. *(Courtesy Accelerated Technologies)*

These operations are highly repetitive and very tedious. Yet they demand attention to detail, good hands, and a good eye. When benching a prototype, technology no longer controls the quality. Instead, it is literally in the hands of the bench technician. Compounding this situation is the fact that these prototypes must be delivered rapidly. Keeping one eye on the details and the other on the clock requires a unique personality. For this reason, it is imperative that part benching be considered early in the implementation phase, not as an afterthought.

Conclusion

While thorough and detailed, the process described above is not complete for any one system. There are many other processes, actions, and considerations for each technology. If the goal is to evaluate, select, and implement a rapid prototyping technology, further investigation is required. A basic understanding of the system-specific process will suffice for the evaluation and selection phase. A detailed understanding is required when it is time to implement the selected rapid prototyping technology. Of course, all of this can be ignored if the course of action is the use of a qualified rapid prototyping service bureau.

PROCESS STRENGTHS AND LIMITATIONS

From the detailed description of the process, many of the strengths and limitations of rapid prototyping are revealed. With numerous methods of prototyping, it is imperative that these factors are weighed when selecting the best method for a prototype, whether for an in-house implementation or the subcontracting of service bureau work.

Strengths

From the name of the technology and the process description, the most significant strength is the most obvious: speed. However, most view the speed advantage only in terms of the time to construct the prototype, ignoring the overall speed advantage of the entire process. With fewer steps in the process, automation, and less dependency on labor and unattended operations, the entire process, from data receipt to prototype delivery, is fast and efficient.

With the additive nature of rapid prototyping, extremely complex designs are constructed with little impact on time and labor. For rapid prototyping, few designs are impossible. While conventional processes may require multiple setups or fabrication of a part in sections, rapid prototyping can build most parts in one operation.

While speed in terms of build time is a powerful advantage, it does not fully account for the overall throughput and operational efficiency. Rapid prototyping gains advantages in these areas from multiple factors. Unattended operations allow rapid prototyping devices to churn out parts 24 hours a day, seven days a week without the addition of a second and third shift. Additionally, the automation of the process requires less labor. In fact, it is common for one individual to prepare data for and operate two, four, or even eight machines.

Adding to the throughput and efficiency advantage is the fact that a single machine run can process dozens of parts. Most other prototyping and manufacturing processes are sequential, addressing a single part at a time. When all of these time elements are combined, rapid prototyping offers the ability to receive data for dozens of parts at 4:00 p.m. and have them available for delivery mid-morning of the following day. Without multiple shifts and multiple machines, this throughput is unavailable for most other processes.

A final strength of rapid prototyping is that the operation is linked directly to the digital data definition. There is no requirement for human intervention, interpretation, or manipulation of the CAD data. This not only adds to the efficiency of RP but also removes a variable from the process. In effect, RP is the direct output of a CAD system. The resulting prototype illustrates exactly what has been designed. The RP device is used as a proofing tool to verify the quality and accuracy of the CAD data, a valuable by-product that can protect a company from downstream problems.

Limitations

Through the process description, the limitations of rapid prototyping become as evident as the strengths. The most obvious limitation is the quality of the prototype.

When compared to CNC machining, quality would be best described as reasonable. With the ability to produce dimensions within ±0.001 in. (±0.03 mm) and smooth surfaces and holes that are perfectly circular, CNC has an advantage over rapid prototyping. While rapid prototyping vendors often advertise dimensional accuracy of ±0.005 in. (±0.13 mm), this is usually a best-case generalization. Realistically, on a feature-by-feature basis, accuracies range from ±0.005 to ±0.030 (±0.13 to ±0.76 mm).

There is also an issue of repeatability. With factors such as part placement, environmental change, and system calibration affecting the output, the same part built on two different days, or two different machines, is likely to yield varying results.

Another consideration is the stair stepping inherent in the additive process. Every rapid prototype will have some degree of stair stepping unless all walls and features are perfectly horizontal or vertical. These stair steps diminish the visual appeal and accuracy unless finishing operations are performed.

The benching required to eliminate stair stepping leads to the next limitation, human involvement. While rapid prototyping is for the most part an automated process, the quality of the prototype and its delivery lead time are not a function of the technology, but rather a function of the work force. The last step in the process, benching, is a labor-intensive operation. The skill of the bench technician and the ability to work rapidly, while maintaining accuracy, are critical to the quality and delivery time of the prototype.

Materials are also a limitation of rapid prototyping. This is true in terms of cost, selection, and material properties. While the industry has come a long way from the early days of stereolithography, when only a single, brittle material was available, the selection still pales in comparison with conventional processes. Although a technology like stereolithography may have dozens of materials from which to choose, many technologies are limited to one or two proprietary materials supplied by the equipment manufacturer. In addition, there is no way to produce a rapid prototype that offers all of the physical material properties of any production-grade plastic or metal. Some systems do offer acrylonitrile-butadiene-styrene (ABS), nylon, or polycarbonate, and the material properties can approach those of production materials, but the way that the materials are processed in the machine yields properties different from those of a machined or injection-molded plastic part.

Strength, Weakness, or Opportunity?

Little in life is truly black and white, and rapid prototyping is no exception. Take, for example, expense. Comparing equipment cost and operating expense with a CNC mill can make rapid prototyping appear to be unreasonably expensive. However, when the total cost of operation is reviewed, the difference diminishes. With staffing, facilities, equipment, throughput, and efficiency considerations, rapid prototyping can, in many cases, be more cost effective than many conventional processes.

Similarly, weakness can be turned into strength. When plastics were first introduced, they were an inferior replacement for metal. Applied to designs and applications that were unthinkable for metal, the weaknesses of plastics were overshadowed by the opportunity. While rapid prototyping materials can be viewed as a limitation, creativity and innovation may lead to unique application of these materials, which cannot be matched by any other material or process. One example is the use of stereolithography for photo-reactive stress testing. While this application has been used in relatively few applications, it is one example of how a perceived weakness can be turned into an opportunity.

The key element in weighing strengths and weaknesses is to review the technology with respect to current and future needs. No one can proclaim that either rapid prototyping or CNC machining is the best for every application. In some cases one will be the obvious choice, and in others it will not be so clear. The important consideration is that the user must select the right technology for the right application.

CHAPTER 4

Classes of Rapid Prototyping Systems

The rapid prototyping industry has yet to reach a level of maturity and stability where each technology can be slotted into a technology class. Unlike automobiles, where sedan, coupe, compact, full size, minivan, wagon, luxury, or economy have very definite meanings, rapid prototyping does not have commonly accepted classifications. For example, some consider all rapid prototyping devices to be 3D printers, and others believe this term has a very specific connotation.

While there is disparity in the definition of the rapid prototyping classes, and a high probability that they will change over time, for clarity purposes, it is important to define the classes of technology. To avoid confusion, it is critical to restate that this text focuses only on additive processes. With the inclusion of subtractive and formative methods of rapid prototyping, it would be virtually impossible to apply any reasonably descriptive terms to any classification system.

There is a great diversity in rapid prototyping systems. Purchase prices range from \$30,000–800,000. Materials range from plastics to metals and from resins to powders. Build envelopes range from 8 in.^{3} (131 cm^3) to well over 8 ft^3 (0.2 m^3). The sizes of the systems range from that of an office copier to that of a five-axis CNC milling machine. Operating environments range from engineering offices to isolated, HVAC-controlled labs.

With the system diversity, rapid prototyping systems could be classified in a number of ways, including process type, process functionality, material type, size, cost, or ease of operation. As time passes, there will be a discrete segmentation of the rapid prototyping market, and this segmentation will occur along the lines of functionality and price. Today, however, there is no clear delineation of technology by class. Rather, most systems cross between several of the technology classes.

Camera Analogy

Classifying the rapid prototyping technologies is not intended to indicate strength, weakness, or limitation. The relative strength of each class is dependent on the application.

In the world of photography, there are $100 point-and-shoot cameras and $5,000 professional cameras. Both deliver the same output, a photograph. For the consumer market, the $100 camera allows ease of use by eliminating all of the image controls that a professional camera offers. Yet, the output quality of the low-end camera is more than suitable for family photos. On the other hand, a professional photographer could not shoot a quality image with a consumer-grade camera. The trade-offs are found not only in the camera cost but also in the knowledge and skill required to shoot the image.

With the rapid prototyping market classification, a similar relationship exists. All of the technologies produce the same output, a prototype (or tool) constructed in an additive fashion. While the 3D printer market offers low cost, easy-to-use devices, the amount of control over the output is minimized for the sake of cost and ease of use. When using an enterprise prototyping center, expense increases, as does the required skill level. In return, there is a greater range in what can be produced.

It is also important to consider that the construction process is not a valid classifier. A technology that uses an inkjet printing technique is not always a 3D printer. For example, the Z Corporation technology uses an inkjet process developed at the Massachusetts Institute of Technology (MIT). However, other MIT-based technologies address rapid tooling and rapid manufacturing.

3D PRINTERS

3D printers are generally considered to be small, affordable, fast, and easy to use. Additionally, fitting with the general perception of a 2D printer, they are shared peripherals or personal devices located in, and operated by the design, engineering, or manufacturing department.

Presently, no systems are a perfect match for the 3D printer classification. Some systems are very close, but improvement is required to satisfy the user's demands for a 3D printer. Most notably, today's technology is often a bit larger than a copy machine and not nearly as clean and silent in its operation. In the future, 3D printers will range in size from that of a desktop printer to a departmental copy machine.

"Affordable" is a relative term, and over time, it will imply a price significantly less than that of today's 3D printers. In the future, 3D printers may have a price in the $500–25,000 range. Today, these low-cost devices range in price from $30,000–70,000.

The 3D printer is expected to be a plug-and-play device that requires little setup, training, or maintenance. Essentially, they are systems that anyone can use—a direct peripheral available to the team rather than a service offered to and managed by specialists. The push-button nature of the technology simplifies the operation but limits the availability of user controls, much like a point-and-shoot camera.

Speed of prototype delivery is another quality of the 3D printer. However, this speed is often not a result of the system's speed, but rather a function of its location and operation. With a 3D printer close by and operated by the department, many time-consuming steps associated with a rapid prototyping lab are eliminated. Therefore, a 3D printer may be slower than an enterprise prototyping center and yet deliver a single prototype in a more expedient fashion.

With the ease-of-use and simplicity of operation, some sacrifice in functionality and capability is generally expected. These come in the form of reduced capacity (size of prototype), diminished quality, and limited material selection.

The 3D printer segment is somewhat analogous to the copy-machine market. For copiers, there are solutions intended for departmental use, which are more expensive, have higher throughput, and are larger than the desktop. Copiers also are designed for home-office use, offering lower cost, smaller size, and fewer pages per minute. In the future, the 3D printer market will offer departmental and personal devices.

The copier and printer analogy help to distinguish 3D printers from enterprise prototyping centers. For large corporations, there may be a department that operates a large, full-function, high-speed copy center. Yet, an engineering department most likely has its own copy machine. The decision on which to use is based on the type of document and the desired quantity. The enterprise prototyping centers are like the devices used in the copy center, and the 3D printers are like the departmental copy machines. Similarly, many documents are printed on desktop inkjet or laser printers. For higher quality and larger volumes, the same document may be sent to a commercial printer. The enterprise prototyping centers are like the printing presses used by commercial printers, and the 3D printers are like desktop printers.

ENTERPRISE PROTOTPYING CENTERS

The enterprise prototyping center category represents the classic rapid prototyping system, the type of device that first gave birth to the industry. These devices excel at producing models, prototypes, and patterns. Devices in this category are multifunctional, robust, and feature rich. These systems also offer a maximum part size that exceeds that of a 3D printer.

Unlike 3D printers, these devices are best suited for a shop floor or lab environment, as shown in *Figure 4-1*. There are a number of reasons, most notably the size of the device. Rapid prototyping equipment in this category will typically have a footprint greater than 5 × 3 ft (1.5 × 0.9 m) and stand over 5 ft (about 1.5 m) tall. Few office environments could accommodate a device of this size. Beyond size, environmental conditions separate these systems from

FIGURE 4-1. Unlike 3D printers, enterprise prototyping centers are best suited for a lab-like environment, isolated from other areas of the facility. *(Courtesy 3D Systems)*

3D printers. These devices are often isolated from other areas. The isolation serves multiple purposes, namely debris and vapor containment, and temperature and humidity control.

Enterprise prototyping centers are tools for a skilled operator, allowing a high degree of flexibility in the parameters used to construct the prototype. With this functionality, the systems require trained, skilled, and experienced operators. Therefore, these devices are likely to be staffed by an individual or team dedicated to the process. These devices are most likely to be operated in a separate department that supports the needs of the design and manufacturing teams and other departments—collectively "the enterprise." Unlike the peripheral concept of the 3D printer, enterprise prototyping centers are implemented as a support service where others execute the entire process to deliver the prototype.

Enterprise prototyping centers often require supporting equipment for the cleaning and benching process, equipment not suited for the office. Examples include ultraviolet (UV) ovens, parts washing tanks, solvent tanks, downdraft benches, and material recovery devices. This leads to another consideration: For the majority of applications, prototypes will demand additional finishing to make them suitable for the intended application. This benching operation will require additional floor space and staff, much like any model shop environment.

While 3D printers are expected to be quick in part generation, enterprise prototyping centers are expected to have high throughput. This is a combination of overall system speed and the capacity for a large number of parts in a single run. These devices often gain operational efficiency from the use of the entire build capacity. For example, they may be most effective when multiple parts are run concurrently, and highly inefficient when a single, small part is built.

The final differentiation between enterprise prototyping centers and 3D printers is the overall cost of ownership. Enterprise prototyping centers typically cost 2–25 times that of a 3D printer. Additionally, the ongoing operational expense is much higher when including labor, floor space, maintenance, and materials. As time passes, this gap in cost will increase due to economies of scale. The potential for sales of 3D printers is much greater than for enterprise prototyping centers. With increased unit sales of 3D printers, costs are amortized over more units, distribution and support becomes less costly, and components will become standardized. The enterprise prototyping center market will not enjoy these same economies of scale.

DIRECT DIGITAL TOOLING

Both rapid tooling and rapid manufacturing applications are too immature to fully define. As time passes and research turns into commercial applications, both categories will mature and develop, most likely in unexpected ways. Yet, a categorization can be applied with the capabilities of current technology.

Direct digital tooling systems address the rapid tooling applications. They share all of the characterizations of the enterprise prototyping centers. Yet, there are additional parameters to apply.

While devices in this category include both direct and indirect solutions, the general perception is that these devices should produce metal, ceramic, or composite tooling. In other words, the rapid prototyping device produces tools or inserts without an intermediate step.

As there are more developments, direct digital tooling devices will become specialized machines. Already, there are devices dedicated to tooling and others that use the advantages of additive processes to deliver tools that can only be produced in this way. Two examples are conformal cooling and gradient materials.

There is a lack of specialized equipment in the direct digital tooling category. Many devices use indirect approaches. An example is the use of stereolithography to produce a plastic pattern used in secondary operations to form the tool or tooling insert.

The direct digital tooling category has several barriers, but it will develop. Barriers include accuracy, material properties, and surface smoothness. The intended benefits of rapid tooling—time and cost reduction—have been an evasive goal. In the past decade, significant advances have been made in CNC machined tooling, advances that impinge on or fully overwhelm the time/cost advantage of direct digital tooling devices.

For some, rapid tooling is synonymous with rapid manufacturing. These individuals view the generation of a production-grade tool as a key component of rapid manufacturing. While there is nothing wrong with this view, this application is not considered in the direct digital tooling classification.

DIRECT DIGITAL MANUFACTURING

High throughput, large capacity, and fully monitored devices are what defines the direct digital manufacturing category. This class of rapid prototyping technology produces finished goods rather than prototypes, patterns, or tooling, and the application of the technology is called rapid manufacturing. In its most eloquent role, this type of device enables the production of a single unit on demand.

While many speculate that industry will ultimately offer personalized, one-of-a-kind products for each consumer, the challenges and logistics of consumer demand will be major obstacles to this goal. The more likely scenario is that direct digital manufacturing devices will allow the production of products in extremely small lot sizes. For manufacturing plants around the world, the advantages are tremendous. These could include:

- reduced inventory expense
- elimination of the obsolescence of inventoried products
- reduced dependency on accurate sales forecasting
- elimination of tooling
- greater flexibility for product change

While it is unlikely that razor blades—produced in the millions per day—will take advantage of rapid manufacturing, there are a great number of products offered in low to moderate quantities. Many products sold to business and industry are produced in annual quantities well under the daily production rate of the razor blade. Aerospace, industrial goods, and medical devices are just a few examples where the annual demand is unlikely to exceed 10,000 units. It is in these applications where direct digital manufacturing will be first applied.

Currently this segment is populated only with crossover devices, like stereolithography and selective laser sintering machines. However, general-purpose devices will not satisfy this category in the future. Instead, devices will be designed, from the ground up, to be a manufacturing tool. The systems will incorporate all of the processes, procedures, and controls that exist on the manufacturing floor. For example, future direct digital manufacturing devices will include feedback and feed-forward monitoring and statistical process control.

Future developments will also lead to a wide range of production-grade materials with high processing speeds. This combination will allow the direct digital manufacturing device to produce a wide range of products with properties suited to consumer products and industrial goods. Throughput rates will increase through process enhancement and larger build capacities. In the next decade, tremendous strides will be made in this area. The interest is high, and the potential demand is huge. A single plant that operates one or two enterprise prototyping centers, or a dozen 3D printers, is likely to require dozens, if not hundreds of these direct digital manufacturing devices.

SYSTEM CLASSIFICATION

To add some depth and clarity to the four classes of rapid prototyping technology, the four processes discussed in this book are categorized.

Stereolithography

The primary class for stereolithography devices is enterprise prototyping centers. With system prices of $180,000–800,000, they are clearly not 3D printers. Additionally, stereolithography has a good balance of speed, quality, materials, and throughput, making it a good technology for a broad range of applications. Stereolithography is commonly applied to form/fit models and patterns. With the right material, limited functional testing also can be performed. 3D Systems' Viper si2™ is shown in *Figure 4-2* and the SLA® 7000 is shown in *Figure 4-3*.

Stereolithography also crosses over into the direct digital tooling class. When the description of direct digital tooling is extended to include indirect methods (pattern-based), the quality of a stereolithography pattern enables the development of rapid tooling. The technology also has had some limited success as a direct tooling generator through its ACES™ injection molding (AIM) process. AIM has been

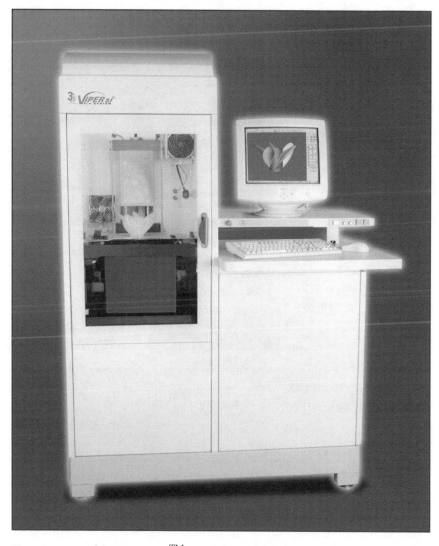

FIGURE 4-2. The Viper si2™ is a stereolithography system with high resolution capabilities and a working area of 10 × 10 × 10 in. (254 × 254 × 254 mm). *(Courtesy 3D Systems)*

applied to extremely short runs of injected molding materials that do not have demanding injection parameters.

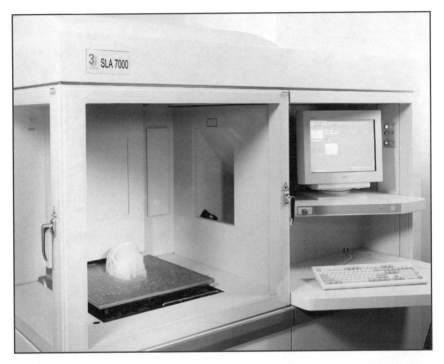

FIGURE 4-3. The SLA® 7000 is the largest and fastest stereolithography device from 3D Systems. It has a working area of 20 × 20 × 24 in. (508 × 508 × 610 mm). *(Courtesy Accelerated Technologies)*

As with direct digital tooling, when indirect approaches are considered, stereolithography can be a direct digital manufacturing device. Two popular examples include the fabrication of components for Formula 1 and NASCAR race cars and the production of clear, disposable orthodontic appliances. Both manufacturing applications use stereolithography as a pattern generation tool for the production of end-use items.

Selective Laser Sintering

The primary class for selective laser sintering devices is enterprise prototyping centers. Selective laser sintering

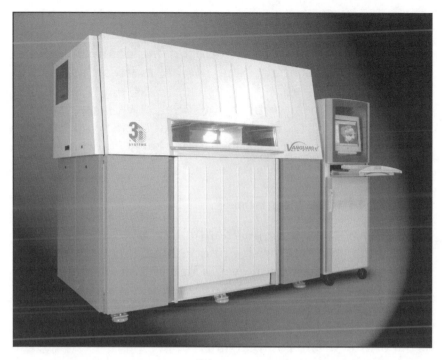

FIGURE 4-4. The Vanguard si2™ is a selective laser sintering device from 3D Systems. The system has a build envelope of 13 × 15 × 17 in. (330 × 381 × 432 mm) and it processes metals and plastics. *(Courtesy 3D Systems)*

devices sell for approximately $300,000. Best known for its material properties, the technology is often applied to functional prototyping applications. These materials include powdered thermoplastics and binder-coated metals. The Vanguard si2™ from 3D Systems is shown in *Figure 4-4*.

Selective laser sintering also crosses over into the direct digital tooling class. With its binder-coated metals, the process can directly produce tooling inserts for applications like injection molding. Of the four technologies discussed, selective laser sintering is the best example of direct digital tooling.

The functionality of materials allows selective laser sintering to cross over into the direct digital manufacturing class. It can be applied to the direct manufacture of production parts. Surprisingly, one of the most stringent and demanding applications, aerospace, was one of the first rapid manufacturing applications. Selective laser sintering has produced parts for the space shuttle and the international space station. Parts also have been incorporated in military fighter aircrafts.

Powder-binder Printing

The primary classes for powder-binder printing devices are 3D printers and enterprise prototyping centers. Built upon inkjet printing methodology, Z Corporation has one of the lowest priced systems on the market, and it has a size that is approximately that of a copy machine. These devices, the ZPrinter™ 310 (*Figure 4-5*) and Z®406, are well positioned as 3D printers. The company also offers a larger device, the Z®810 (*Figure 4-6*). While it uses the same inkjet process, the physical size and intended application position it as an enterprise prototyping center.

The Z Corporation technology has been applied to functional prototypes and patterns. With a reasonable build envelope and high throughput, the 3D printers are often positioned as enterprise prototyping centers that compete with stereolithography, selective laser sintering, and fused deposition modeling. Furthermore, the pattern and functional prototype applications demand additional part finishing, operations that are not suitable for the office environment.

Some applications of the Z Corporation technology are best classified as direct digital tooling applications. Most recently, the company has introduced materials and a process for the construction of molds for the casting of metal. Using a plaster-based material, the technology prints a mold for each metal casting.

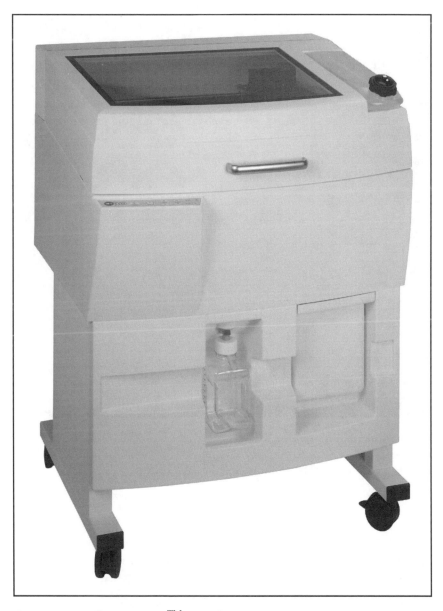

FIGURE 4-5. The ZPrinter™ 310 from Z Corporation is a 3D printer sized for an office environment. This system has a build capacity of 8 × 10 × 8 in. (203 × 254 × 203 mm). *(Courtesy Z Corporation)*

FIGURE 4-6. Z Corporation's largest system, the Z®810, incorporates a powder-feed system to supply the 20 × 24 × 16-in. (508 ×610 × 406-mm) build chamber. *(Courtesy Z Corporation)*

Fused Deposition Modeling

The primary class for fused deposition modeling devices is enterprise prototyping centers. The technology line-up for fused deposition modeling includes systems that range in $150,000–250,000 (*Figures 4-7* and *4-8*). Offering acrylonitrile-butadiene-styrene (ABS), polycarbonate, poly-phenylsulfone, and other materials, the technology is

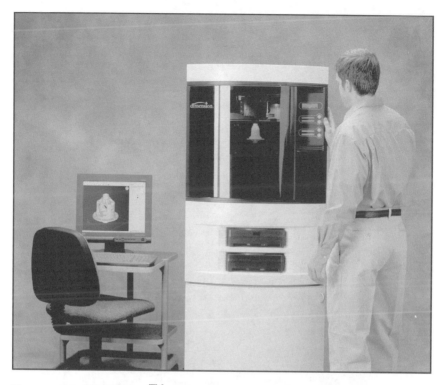

FIGURE 4-7. Dimension™, a 3D printer from Stratasys, produces prototypes in acrylonitrile-butadiene-styrene (ABS) that can be as large as 8 × 8 × 12 in. (203 × 203 × 305 mm). *(Courtesy Stratasys, Inc.)*

commonly used for functional prototypes. Also suited for concept and form and fit models, fused deposition modeling addresses the full scope of engineering applications.

The fused deposition technology has a device that fits the characteristics of a 3D printer: the Dimension™ from Stratasys, Inc. However, Stratasys does not classify it as a fused deposition modeling (FDM) system. Although it uses the same extrusion process, the company separates Dimension from its FDM product line. The Dimension™, with a list price of $29,900, is sized to fit in an office environment and offers the simplicity of operation expected from a 3D printer.

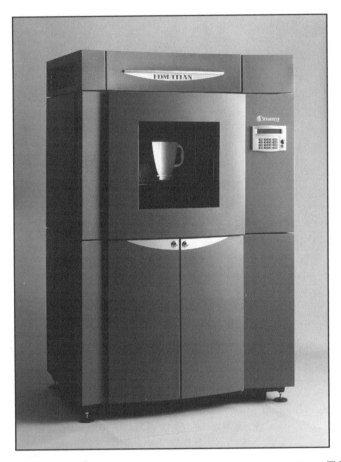

FIGURE 4-8. The fused deposition modeling machine, Titan™, has a build envelope of 16 × 14 × 16 in. (406 × 356 × 406 mm). The system uses a variety of thermoplastics including ABS, polycarbonate, and polyphenolsulfone. *(Courtesy Stratasys, Inc.)*

Like selective laser sintering, the functionality of the materials enables fused deposition modeling to serve as a direct digital manufacturing solution. Fused deposition modeling parts have been used as replacement parts for production line equipment and as a short run solution for military hardware.

SPECIALIZATION REPLACES GENERAL PURPOSE

Rapid prototyping is a young industry, and therefore it is not surprising that the listed technologies have broad application over multiple classifications. Lacking the maturity of an established industry, companies offer general-purpose devices as they address new, emerging applications.

The industry has not been ready for specialized systems that address specific applications. The market has demanded that rapid prototyping technologies address a large portion of the design and manufacturing process. In light of moderate growth in the unit sales of equipment, rapid prototyping vendors have responded with developments that extend a given product to a broader base of applications. This positions many rapid prototyping systems as general-purpose devices.

As the rapid prototyping industry matures, customers will begin to demand technologies designed and developed for specific applications. Many would agree that a system, tool, or device that has many purposes rarely does them all well. Instead, the system performs all functions reasonably. As greater demands are placed on rapid prototyping and its output, the industry will respond with specialized systems.

The class of 3D printers has already begun the change from general-purpose to specialized application devices. They address the specific need for fast, easy to use, and affordable prototyping devices for the early conceptualization of a design. The leading systems do not attempt to address the higher-end applications of functional prototypes and tooling patterns. The next wave of specialization will likely be in the area of direct digital manufacturing. For those who have applied enterprise prototyping centers to direct digital manufacturing, there is the realization that the general-purpose device is not entirely satisfactory for the demands of production. Once market growth is evident, system manufacturers will respond with specialized devices that address the specific needs of the manufacturing and production environment.

As specialization replaces general application, the classes of rapid prototyping technology will be clearly delineated and fully defined. While 3D printers, enterprise prototyping centers, direct digital tooling and direct digital manufacturing may not be the resulting classifications, they are appropriate to the industries and applications currently served. Over time, it is likely that most, or all, of these classes will be redefined.

If rapid prototyping is analogous to the 2D printing market, the future will also lead to a host of subclassifications. For the 2D printing, the desktop printer market is subdivided into inkjet and laser printers. For the commercial printing market, technology is further classified by output. For example, there are low-volume, two-color printing presses and high-volume, four-color, web-fed printers. The rapid prototyping industry will follow suit. Each category will be subdivided into subclasses defined along the lines of output and capability. There will be a clarity that is much like that offered when describing a computer as handheld, notebook, desktop, or server.

CHAPTER 5

Applications and Benefits

APPLICATIONS

While rapid prototyping is often viewed as a design and engineering tool, the application of the technology is much broader. Since the introduction of the technology, rapid prototyping applications have expanded to cover the entire spectrum of business operations. Design, manufacturing, sales and marketing, and executive management are some of the corporate departments that use and benefit from rapid prototyping.

Design

Rapid prototyping is applied in every phase of the design process, from initial concept development to final design and release to manufacturing. While there are many underlying benefits in each application, both personal and corporate, the primary value derived is that it is a quick way to provide for visualization and understanding of a design and its function. Making the design real and tangible fosters improved communication, better decisions, and early error detection.

The applications for rapid prototyping, and the benefits derived from its use, are not unique. There are numerous methods for the development of prototypes. These range from handcrafted models to prototyping tooling for casting and molding. The advantage found with rapid prototyping is it removes barriers that can impede the use of prototypes in the design cycle. With the inherent speed and simplicity of rapid prototyping, companies can prototype early and often, all while accelerating the design of better products.

Concept Models

Very early in the design process, a product will begin as a series of ideas, thoughts, concepts, and rough sketches. As these concepts are narrowed to a few viable designs, they begin to take shape as they are modeled in a computer-aided design (CAD) system. At this stage, the designs take form with a loose description of overall style, size, and form.

To select viable concepts, the design team, marketing staff, and management team review each design. Many organizations also present the concepts to their potential customers. While 3D CAD solid modeling packages can create realistic renderings for the initial review, as shown in *Figure 5-1*, they do not clearly communicate the design, especially to non-technical individuals. To bridge this communication gap, rapid prototyping creates physical models that the members of the design team and those throughout the organization can review. These concept models foster better and faster decisions with the clarity of a real, tangible representation of the design. *Figure 5-2* shows a concept model of a cell phone housing.

During the concept-modeling phase, the rapid prototypes may have a useful life of only a few minutes. With input from the reviewers, the design is modified in CAD, and new concept models are constructed. This cycle of design and prototyping may be repeated many times over a one or two-day period.

For organizations with team members spread across the globe, rapid prototyping is often the first action item following the conclusion of a collaboration session. Here the concept models, representing each design option, are sent to the team members. In this role, some consider rapid prototyping to be the 3D equivalent of a fax machine.

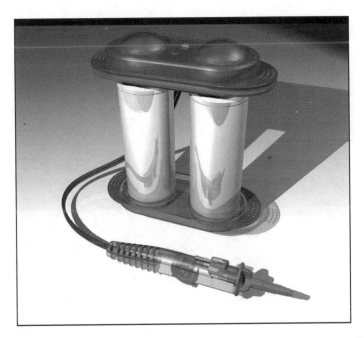

FIGURE 5-1. Realistic images of new products can be generated from CAD tools, but they often do not fully communicate all aspects of a design. *(Courtesy Leyshon Miller Industries)*

Concept Modeling Application

A package design company, specializing in bottles for the perfume and cologne market, was tasked with the creation of a highly stylized bottle for a new, upscale fragrance. With input from the client's marketing team, the company designed several concepts. These were presented as computer-based, photo-realistic renderings to the client. One design was selected and it progressed through tooling. Upon presenting the first samples from the tools, the client's marketing team was shocked to find the bottle was nearly twice the size they had envisioned. With the use of concept models, the situation would have been avoided. The concept models, held in the hand of the marketing team, would have clearly demonstrated the actual size of the package.

FIGURE 5-2. Concept models, such as these cell phone housings, are valuable tools in the iterative process of the early design cycle. *(Courtesy Z Corporation)*

Form and Fit Checking

Once a concept has been selected, the loosely defined product proceeds further into the design cycle. At this stage, all details of the product are defined with consideration of form, fit, function, and manufacturability.

After the detailed design is completed, rapid prototypes are constructed and assembled, as shown in *Figure 5-3*. The fully assembled model, which often includes all internal mechanical and electrical components, clearly indicates any errors in the design and areas in need of improvement. In effect, the rapid prototypes become a proofing tool that serves to detect design flaws early, when they are least expensive to correct.

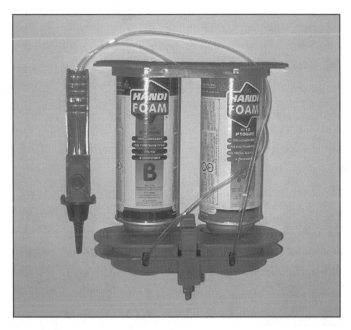

FIGURE 5-3. Fully assembled prototypes of the foam-dispensing unit shown in *Figure 5-1* allow a complete review of all aspects of form and fit. *(Courtesy Leyshon Miller Industries)*

The rapid prototypes also are used to determine the impact of the design and its features on the manufacturing process. With the prototypes, engineering and manufacturing can collaborate on design modifications to improve the manufacturability of the product in terms of cost, time, and quality.

As with concept models, the design and prototyping cycle is repeated multiple times. However, for form and fit analysis the cycle is longer than that of concept modeling due to the amount of analysis, calculation, and evaluation required. Evaluating all product features and reflecting the changes in a CAD file requires more thought and effort than that needed for simplified concept models. In a typical project, the design and prototyping cycle will be repeated multiple times over the course of several weeks.

3D CAD Limitation

Offering excellent tools to analyze a design, 3D CAD includes features for interference checking. However, these software tools do not entirely replace the human element. For a manufacturer of power tools, the design of the main housing looked good in CAD and it passed all checks. Upon receiving rapid prototypes of the design, two flaws that the computer could not detect were found. The first was in the area of the boundary mating the two parts. They did not seat properly. The rapid prototypes illustrated that the joint had no clearance between the mating rib and slot. This was the result of a Boolean subtraction in the CAD system. The second was in the area of mounting standoffs for the motor mount. While the bosses did not interfere with one another in CAD, they did not allow enough space for the motor's mounting flange.

Ergonomic Studies

Ergonomic evaluation may be conducted concurrently with the form and fit review. Ergonomics is a subtle science that measures the interaction of the user with the product.

For ergonomic studies, a physical model is a necessity. It is nearly impossible, at least with today's technology, to predict the feel, comfort, and ease-of-use of a product with digital models. Size, weight, balance, accessibility, comfort, and ease-of-use are all subjective variables. To determine whether a product satisfies the elements that define the user experience, a representation must be put into the hands of people. This requires a physical prototype. As with form and fit analysis, rapid prototypes are constructed from the 3D CAD data and assembled. With the physical models, people can interact with the product and comment on necessary modifications to improve the product and customer satisfaction.

Consumer Testing

A manufacturer of resealable containers had designed a new product line. The company had a clear definition of the force, which consumers found comfortable, to open or close the container. However, it was impossible to measure these forces from the CAD data for the design. To verify this ergonomic aspect, rapid prototypes were created and used as patterns for rubber molding. Using cast urethanes that mimicked the mechanical properties of the intended production material, the sealing and opening forces were measured in an unscientific way. The prototypes were put into the hands of potential customers who were asked to comment on the product. In the first round of design, the consumers reported that it was too hard to open the container. This yielded several more rounds of design and prototyping to achieve just the right amount of force.

Functional Testing

Once form, fit, and feel have been reviewed, the next task is to determine whether the product works. For this testing, rapid prototypes are constructed for functional analysis (see *Figures 5-4* and *5-5*). The goal is to confirm that the product will perform as specified, in the anticipated conditions, without failure.

As functional models and tooling patterns, rapid prototypes expedite the testing process. Determining failure points or performance standards prior to an investment in tooling saves time, reduces costs, and improves the product. Rather than invest time and money in an unproven design, many apply rapid prototyping to detect errors and flaws so they can be corrected efficiently and cost effectively.

FIGURE 5-4. This coffee maker carafe constructed entirely of rapid prototypes (with the exception of the metal band) is functionally tested as hot coffee is brewed, warmed, and served. *(Courtesy Stratasys, Inc.)*

Lesson Learned: Model All Components

A manufacturer of kitchen appliances requested functional models for a new beverage maker. While the function did not include heating the beverage—the rapid prototyping material could not withstand the temperature—all other functional aspects were evaluated for the beverage maker. However, having designed numerous carafes, the manufacturer failed to prototype the beverage container. While the functional testing of the beverage maker allowed the company to modify the design for function and success, the carafe proved to have a problem. Without a functional test, the manufacturer failed to determine that the handle was too close to the body of the carafe. The result was that the consumer's knuckles contacted the heated carafe, resulting in burns. With a functional model of all the components, expensive retooling would have been avoided, and customer satisfaction would have been greatly improved.

Requesting Price Quotes

One underutilized application of rapid prototyping is in the price quotation process. For those who have used the physical model when requesting a manufacturer's quote for production work, the results are sometimes amazing.

Quoting parts, projects and programs demands objective and subjective analysis. For production parts, the quote will include tool design, tool making, material consumption, workflow, assembly, packing, and distribution. In effect, the quote should reflect the actual demands of a complex process that will not be completely defined until an order is received. Taking in account all of these processes and calculating a competitive, yet profitable price

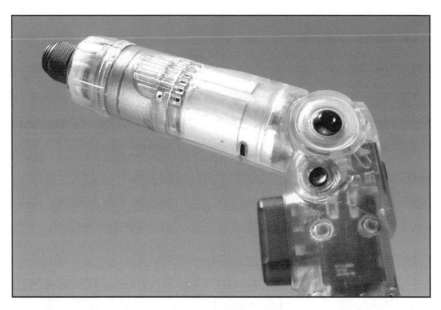

FIGURE 5-5. This fully functional pivoting drill offers the benefits of field-testing and visualization of the inner mechanisms. *(Courtesy DSM Somos)*

is challenging. To err on the side of caution, additional cost allowances are applied when an aspect of the design is confusing or questionable.

The ability of rapid prototyping to clearly describe the product's design allows a full, comprehensive understanding of the proposed work. With this information, assumptions and questions are eliminated from the quoting process. The rapid prototype yields an accurate, finely tuned quote that generates desired profits. With only engineering drawings or CAD data, assumptions and questions result in allowances to cover the worst-case scenario, thereby driving up the price.

The Value of Rapid Prototypes in the Quotation Process

More than a decade ago, an automotive manufacturer tested the use of rapid prototypes for quote requests to determine the value of the effort. In the study, half of the vendors received only engineering prints or CAD files. The others received the same information and a rapid prototype. Amazingly, the half who received the prototype returned a price that was, on average, 25% less than those who did not have the physical model. While the part cost less than a dollar, when building hundreds of thousands of vehicles, the total cost savings far exceeded $100,000 for the production run.

Propsals and Presentations

As anyone who has presented a product to major retailers like Wal-Mart or Home Depot knows, never come to the sales presentation without the product or its prototype.

With only words and pretty pictures, these behemoths will quickly escort a salesperson to the door. When considering a new product line, buyers want to see and hold a "real" product, and rapid prototyping is one method to deliver what they want. *Figure 5-6* shows a rapid prototype that has been finished and painted to resemble a final product.

While not everyone sells to giant retailers, all sales and marketing professionals understand the value of showing and demonstrating a product to get the customer's interest. Often, to ensure sales upon the product launch, companies will approach the prospective customers before the product is manufactured. In these applications, the rapid prototype,

FIGURE **5-6.** Rapid prototypes are often finished and painted to resemble the final product. *(Courtesy Accelerated Technologies)*

fully dressed to look like the real product, is used as a demonstration, illustration, and proposal tool. Companies also use rapid prototypes to inform and excite their sales force prior to the product launch. For marketing purposes, the rapid prototypes are photographed for use in advertising and sales literature, as shown in *Figure 5-7*.

New Product Introduction Tool

A manufacturer of large industrial valves, often measuring 2 ft (0.6 m) or more in height, was on the verge of launching a new product with exciting new features. Rather than waiting for the production valve to be available, the company approached key clients with the product in advance of production. Using scale models constructed from rapid prototypes, the company traveled the world to introduce the new product line. The result was one of the most successful product launches in the company's history.

CAD Data Verification

As a design is finalized and the product approaches release to manufacturing, the integrity of the CAD data is important to ensure a smooth, problem-free process. With today's technology, many products are manufactured directly from the CAD data, but what if there are problems with this file? Defects in the CAD model can have a major impact on downstream processes, possibly resulting in painful delays and additional cost.

Since rapid prototypes are created directly from the original CAD data, without any human interpretation or manipulation, they are often relied on as a proofing tool to verify the quality of the CAD file. Sometimes defects should be obvious to the CAD operator, but in many cases, they remain undetectable unless someone looks for the specific error. The rapid prototype is a direct representation of the CAD data, and therefore offers an accurate reflection of its quality.

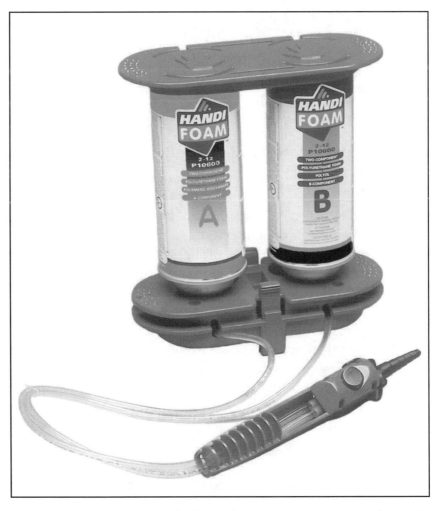

FIGURE 5-7. This image of the foam-dispensing unit was used in early sales literature. It was prepared before finished goods were available. *(Courtesy Leyshon Miller Industries)*

Mistake Proofing

Simple mistakes are often hard to detect. For one manufacturer, a rapid prototype immediately illustrated a design flaw when the outer rim of a component fell off. Upon investigation, it was determined that a very small (less than 0.002 in. [0.05 mm]) gap between the flange and the body was depicted in the CAD data. For another manufacturer, the application of rapid prototyping would have eliminated the scrapping of a tool. Going directly from design to tooling, no one detected that the CAD file had been scaled to twice its intended size. The first injection molding samples arrived, and the client was shocked to find an oversized part. Failure to proof the CAD data for mistakes caused the company to invest more time and money to create a new tool.

Design for Manufacturability Analysis

Every design decision can have a major impact on the cost, time, and effort required to manufacture a part. These design decisions can also affect the quality of the end product. To engage design engineering and manufacturing engineering, two disciplines often working independently, companies have found that rapid prototyping is an excellent collaboration tool.

While collaboration tools are becoming more prominent, in most companies, designs are still "thrown over the wall." When design engineering has done its job, they turn the project over to manufacturing and let them figure out how to make it. This scenario is complicated by the fact that many design engineers are not taught manufacturing processes, and they have little idea of the implications of a design specification. For example, inappropriately specifying a tight dimensional tolerance has a dramatic effect on the manufacturing processes and the ultimate cost.

To tear down the wall, many progressive companies rely on rapid prototypes as a communication tool to help the design and manufacturing teams improve the product's manufacturablity. Like a concept model, which provides visualization of a product and a basis for discussion, the rapid prototype allows designers and toolmakers to fully appreciate a design, its consequences, and challenges. With this collaboration tool, both sides can collaborate and clearly communicate intent and impact. As a result, the company produces better products with lower manufacturing costs.

Patterns

At some point in a product's design cycle, the need for prototypes increases. Commonly there is a need for dozens, or even hundreds of prototypes. However, in most cases, rapid prototyping is a poor solution for large quantities of parts or for a close approximation of the production material's properties. As a result, rapid prototyping is used to fabricate patterns.

A pattern, also called a master pattern, is a form onto which material is poured, pressed, or cast. The pattern is a positive representation of the part from which a negative impression is formed in the tool. These tools are then used to cast or mold multiple prototypes of the component. A common application of rapid prototyping for pattern making is in the rubber molding process. Liquid rubber is poured onto the pattern. Once cured, the rubber mold can produce 1–50 cast urethane prototypes. *Figure 5-8* shows automotive fascias cast from rubber molds. While rubber molding is one of the more common applications, rapid prototyping patterns are applied to numerous processes including investment casting, epoxy tooling, cast metal tooling, plaster mold casting, and vacuum forming.

Beyond Design Engineering

Rapid prototyping applications are not limited to the conventional role of prototypes for the design process.

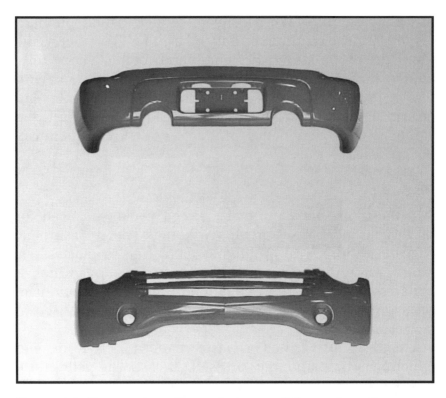

FIGURE 5-8. For an automotive project, stereolithography patterns were used to create rubber molds for urethane castings of these fascias. *(Courtesy Auburn Engineering)*

Since the early 1990s, research and development in other applications and disciplines has led to exciting opportunities. Some industry experts believe that in the coming years the application of rapid prototyping to nontraditional roles will far exceed that of the model, prototype, or pattern.

Rapid Tooling

In the early to mid-1990s, interest swelled in the area of rapid tooling. As designers and toolmakers realize, the time and cost for prototype tooling is a burden on any project. When a design nears or reaches finalization, there is a demand for production-like parts in moderate to high quantities. The applications are numerous: full functional testing, sales samples, manufacturing line testing, and consumer focus groups, to name a few. Some also push beyond prototype tooling to a "bridge-to-production" solution. This bridging solution allows the early molds or dies to serve as both a prototype tool and a production tool. In effect, the rapid tool produces production parts in the period between the launch of production tooling and the receipt of finished goods.

When applying rapid prototyping to tooling solutions, the goal is to decrease the time to produce the tool while simultaneously decreasing cost. While machined tooling can range from $5,000–100,000 and take four to 12 weeks, rapid tooling offers the promise of slashing both of these factors in half (or more). For a detailed discussion of rapid tooling, see Chapter 10. Examples of rapid tooling are shown in *Figures 5-9* and *5-10*.

Rapid Manufacturing

When the time comes to manufacture a product, the investment in tooling can be quite substantial; costs of $50,000–250,000 per tool are common. Moreover, it is a time-consuming process. A production-grade tool can easily take months to produce. The concept of rapid manufacturing is the elimination of the high cost and long delays associated with tooling. In rapid manufacturing applications, the rapid prototyping device produces, in small quantities, the product that the consumer purchases.

Innovative companies have applied rapid manufacturing, as illustrated by the camera mount in *Figure 5-11*. However, it is not a common application for most organizations. At present, rapid manufacturing is a dream of many and a potential threat to others. In the coming years, rapid manufacturing will become commonplace. But significant developments need to occur before the application takes root.

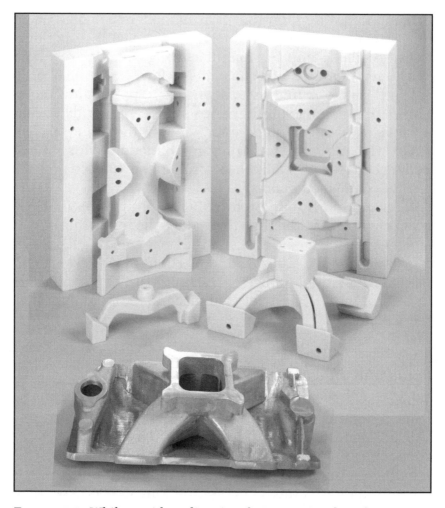

FIGURE 5-9. While rapid tooling is often associated with injection molding, it also applies to other manufacturing processes, as illustrated by this sandcasting mold for the production of a metal casting. *(Courtesy Z Corporation)*

For more information on rapid manufacturing, see Chapter 10.

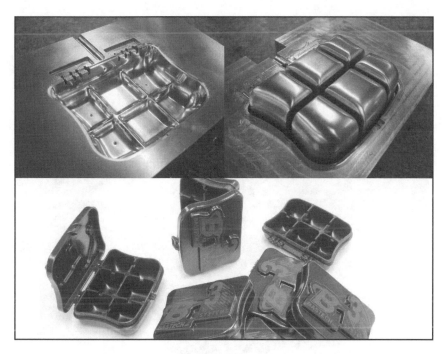

FIGURE 5-10. These injection-molded boxes were produced from rapid tooling constructed with selective laser sintering. *(Courtesy Bastech, Inc.)*

Nontraditional Applications

Some believe the true power of rapid prototyping will be realized in applications that have nothing to do with the design and manufacture of consumer and industrial products. There is evidence that this will prove true in the not too distant future.

Current and future applications of rapid prototyping in the fields of science and medicine will include:

- biomedical modeling to recreate human bone structures and internal organs for review, study, and analysis (see *Figure 5-12*)
- molecular modeling to grow representations of DNA, bacteria, and chemical compounds (see *Figure 5-13*)

FIGURE 5-11. This camera mount for military vehicles is just one example of rapid manufacturing's potential. *(Courtesy Stratasys, Inc.)*

- bone scaffolding to grow replacement bone structures that incorporate beneficial chemical compounds to stimulate rapid regeneration (see *Figure 5-14*)
- producing custom dosages and blends of prescription medicines for individual patients
- custom limb replacements

In building construction and architecture, current and future applications of rapid prototyping include:

- design modeling to replace or augment blueprints and CAD renderings (see *Figure 5-15*)

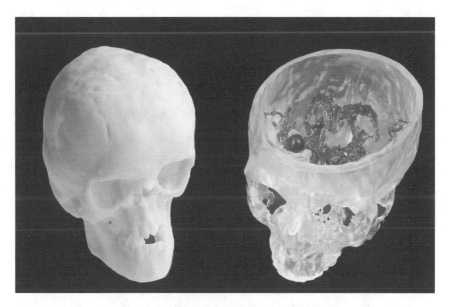

FIGURE 5-12. Increasingly surgeons are using rapid prototyping for pre-surgical planning. These skulls, produced by powder-binder printing (left) and stereolithography (right), offer a visualization tool of the interior and exterior structures. *(Courtesy Medical Modeling, LLC)*

- growing structural elements of a home or office building, allowing increased design flexibility and reduction in construction time

Current and future applications of rapid prototyping in the fields of archaeology, paleontology, and forensic science include:

- growing models of human remains for research and display (see *Figure 5-16*)
- growing a model of an unidentified victim's bone structure to serve as the base for forensic modeling and reconstruction
- scanning and growing complete skeletons of creatures of the past for research and display

Research is underway for the application of rapid prototyping to micro-electromechanical systems (MEMS). These

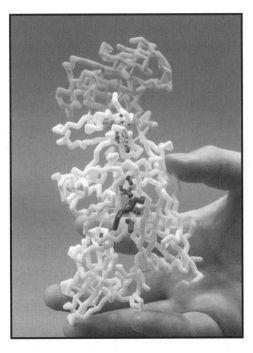

FIGURE 5-13. An elegant example of complex geometry construction, the molecular model of a protective antigen is used as a research and learning aid. *(Courtesy Center for BioMolecular Modeling, Milwaukee School of Engineering)*

micro-machines, measured in microns, approach the size of a single red blood cell.

Documenting and detailing all of the traditional and nontraditional applications that have been conceived would demand books of their own. The previously listed applications are offered to create excitement for the technology and to serve as fuel for creative and innovative applications as personal and unique as the products a company designs.

The applications for rapid prototyping are virtually limitless. With removal of the constraints of conventional processes, amazing things can be achieved. When combined with the inherent speed and simple sophistica-

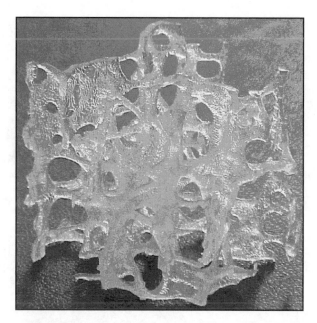

FIGURE 5-14. For verification of computer simulations of the affects of osteoporosis, a model of bone structure was tested. *(Courtesy CADCAMNet.com)*

tion of the process, all types of organizations can realize tremendous and powerful benefits.

BENEFITS

The benefits of rapid prototyping are as diverse as the potential applications. They can be unique to a company, individual, project, or part. As the application base expands, so will the benefits derived from the technology. Effectively, like the applications, the benefits of rapid prototyping are limited by only imagination and innovation. Yet, many of the benefits of rapid prototyping can be achieved through any number of technologies and processes. It is quite possible that a skilled model maker can handcraft a prototype that delivers all of the same

FIGURE 5-15. Architects, like engineers, use rapid prototypes to clearly communicate aspects of the design that may be misunderstood when viewing CAD data or blueprints. *(Courtesy LGM)*

benefits. It is also possible that computer numerical control (CNC) machining can deliver the same or even greater benefits. All prototypes communicate a design, and all patterns create tools; the differences arise when rapid prototyping is applied to the right application.

FIGURE 5-16. When archaeologists discover artifacts or human remains, rapid prototyping can be used to produce research and display models, as in the case of this human skull from the 1600s. *(Courtesy Accelerated Technologies)*

As with other prototyping methodologies, rapid proto-typing allows a product to take physical form. When it does, all of the complexities and subtle nuances of the design are clearly communicated to people of all educational levels, backgrounds, and disciplines. It is this clear communication that stops errors and flaws from making their way into downstream applications. The key difference between rapid prototyping and all other methodologies is that this technology creates a physical communication tool quickly, even for the most complex designs.

Spurred by rapid prototyping, quick, clear, and concise communication offers tremendous value to any company.

With a clear design representation, errors are detected early, when there is still time to fix them with little or no impact on the project's time or cost. Clear representation of a new product design also can be a powerful promotion tool, resulting in increased sales. Communication of an intended design can allow manufacturing engineers the opportunity to simplify a design to reduce manufacturing expense and time, while creating a higher quality product that the consumer enjoys.

Whether rapid prototyping is applied to models, prototypes, patterns, or tools, it enables a clear understanding of all aspects of the product and its design. Thus, when compared to all other processes, the primary benefit of rapid prototyping is that it offers quick, clear representation of even the most complex designs, which fosters clear communication.

Another key benefit of rapid prototyping, and one rarely discussed, is that the technology frees designers and manufacturers from the constraints of subtractive and formative technologies. When machining or molding parts or prototypes, there are considerations that affect the design and limit the possibilities. With the additive nature of rapid prototyping, these constraints are removed. This liberation promotes creativity and innovation in design and manufacturing.

From the aforementioned two general benefits stem the balance of the value that rapid prototyping offers. These can be categorized in the areas of time, cost, quality, and capability.

Time

With the speed of rapid prototyping for complex designs, the delivery time of a prototype or pattern can be greatly reduced. What once took weeks can now be completed in a couple of days. While many focus on the time aspect of the physical construction, rapid prototyping is a quick process from start to finish. Combining the simplicity of the process, unattended operations, and reduction in labor dependency, the entire rapid prototyping process is extremely efficient.

Rapid prototyping operates directly from the digital definition constructed in a CAD file. Although there are operations between CAD data receipt and the start of a rapid prototyping machine run, each is automated and requires little human intervention. For most prototypes, the generation of an STL file and the preparation of build files can be completed in less than one hour. Adding to this efficiency is the ability for rapid prototyping to construct nearly any geometry without modifications. This eliminates the time-consuming effort of modifying part geometries and CAD data to satisfy the requirements of the process. With advances in rapid prototyping, namely the 3D printer market, some systems offer push-button operation for fully automated file processing and job queuing.

When operated within or close to the design and engineering department, 3D printers also offer time advantages. Rather than sending a prototyping request to another department, where it is reviewed, scheduled, and queued behind previous job orders, the 3D printer is operated by the individual who has need for the prototype. This level of control offers operational efficiency that may be unavailable from a departmental operation that is backlogged with work.

Rapid prototyping constructs prototypes quickly, independent of the complexity of the design. Other time advantages result from around-the-clock operation, unattended operation, and multi-piece builds.

Unlike any other prototyping technology or process, the time for constructing a model is solely a function of the physical size. Rapid prototyping devices are dependent on the height of the prototype, the volume of material in the prototype, or both. However, devices are insensitive to the physical design. The time required for conventional prototyping is most often a function of the design, in terms of both complexity and overall size.

With the exception of a multi-cavity tool for a molding process, conventional prototyping is sequential. When one prototype is complete, the next prototype is started. Rapid prototyping is a concurrent operation. Limited by only the capacity of the device, rapid prototyping can build a large quantity of parts, in any combination, in a single machine run. In addition, most rapid prototyping technologies gain efficiencies when multiple parts are concurrently constructed. While a single prototype may take five hours, two of the same prototype may take only seven hours. If a conventional process were able to construct a widget in 10 hours, two widgets could be constructed in 20 hours in two separate operations. If rapid prototyping could build the same widget in five hours, two could be built in seven hours. Therefore, the cumulative difference between rapid prototyping and the conventional process would be five hours (a 50% reduction) for one prototype and 13 hours for two (a 65% reduction). As quantities grow, the difference in time will continue to increase.

Rapid prototyping systems are designed for unattended, 24-hour-a-day operation. This offers significantly more capacity and throughput when compared to conventional processes that demand an operator or operational oversight. There are 8,760 hours in a year. With weekends, holidays, and single-shift operations, manned processes are only available 2,000 hours a year. With nearly 300% more working hours, rapid prototyping offers significantly greater throughput capacity, which in turn decreases delivery time.

Cost

While rapid prototyping can be, in some cases, a less expensive prototyping tool than machining or fabricating, this is not the major cost benefit. Even if rapid prototyping proves to be more expensive on a per-part basis, it still has a tremendous positive impact on production costs and the cost (and therefore profitability) of the product.

With the inherent speed of rapid prototyping, an obstacle is removed. Prior to rapid prototyping, it was often difficult to justify the benefit of the prototype in light of the delays to the design process. When a prototype takes weeks, it can be difficult to see that the potential benefit of the prototype outweighs the impact on the schedule. Since rapid proto-typing shortens the lead time to days, the use of prototypes is much more feasible. When prototyping is used early and frequently, companies can dramatically reduce costs.

With a prototype, design errors can be detected and corrected early. In doing so, there is a decrease in the potential for a poor design to make it to production tooling. This eliminates the cost of tool rework and rebuilding. As previously discussed, rapid prototypes can also decrease product cost when presented as a visual aid for vendor quoting. For some, rapid prototyping implementation can be fully justified within a year due only to the calculation of cost avoidance.

With the availability of the physical model to proof a design, delays in time to market can be avoided. This is true due to time reduction and error avoidance. Some research shows that a project that is under budget but delayed in product launch actually costs a company significantly more than a project that is over budget but on time. The reason is simple: for every day that a product is not in the market, the sales potential is lost forever and that means substantial opportunity costs.

A final cost consideration is that when used effectively, a rapid prototype can spawn cost-saving production ideas. With a rapid prototype as a collaboration tool for both design and manufacturing, simple, yet powerful changes are often revealed. And these changes, such as combining two parts into one or eliminating an undercut, can reduce manufacturing expenses by tens of thousands of dollars. While simplification is beneficial to making cost-cutting changes in tooling, the benefit grows substantially when the production cost of a part is decreased. If a plant spits out 100,000 parts a day, and rapid prototyping helps to identify a $0.01 per part reduction, the result would be a savings of $1,000 a day or $250,000 a year.

By breaking the time barrier and allowing more prototypes to be produced without delaying the design process, the benefits of clear communication for error detection and product improvement translate to significant cost reductions.

Quality

While there are some quality advantages to the prototype itself, for the most part, accuracy, surface finish, and other quality parameters are sacrificed when selecting rapid prototyping. However, quality of the prototype is not the primary issue. Rather, the bigger issue is the quality of the end product and the quality of the departmental output.

As previously discussed, rapid prototyping eliminates barriers to the use of prototypes and, therefore, is a tool that aids in error detection and product improvement. Relying on 3D CAD alone during the prototyping phase limits the interaction with the prototype to a 2D digital representation. When constrained to this environment, sensory feedback is limited. This can result in oversights, errors, and omissions. With the physical prototype in hand, a whole new level of feedback is offered. This physical, tangible form often makes mistakes existing in CAD quite visible and hard to overlook. Thus, rapid prototyping can improve quality within the design department by merely offering clear feedback on a product's design.

Since the majority of company employees and customers are not technically oriented, especially in mechanical design, the rapid prototype offers an opportunity to solicit their feedback. When confronted with engineering drawings or CAD images, understanding of a design is impaired for nontechnical personnel. In the best-case scenario, the nontechnical personnel will ask for further clarification. In the worst case, they will offer input on a design—or approval of it—based on faulty assumptions and conclusions.

For organizations that embrace collaboration between design and manufacturing, a rapid prototype may improve product quality by improving or simplifying manufacturing processes. In design, it is easy to create a product definition that requires manufacturing techniques that introduce process variables. In some cases, these variables may be difficult to predict, control, and improve. As with the use of the rapid prototype to simplify a product design, it can also be used as a discussion tool to simplify the product's manufacture so that highly reliable, repeatable, and controllable processes and techniques may be used. In the end, this translates directly to the quality of the product.

Capability

A key to the benefits previously discussed is that rapid prototyping is insensitive to complexity. Eliminating the need for material removal or material molding and casting, the additive nature of rapid prototyping proves to be efficient and accommodating. No matter how complicated or challenging the design, rapid prototyping can produce it quickly and cost effectively.

With fewer process limitations, rapid prototyping promotes creativity and innovation. For parts and tools, rapid prototyping allows companies to experiment and try new approaches that were previously unthinkable or impossible. In effect, rapid prototyping can be a catalyst to major product and procedural changes.

The most significant changes may flow from the developing applications of rapid tooling and rapid manufacturing. When the capabilities of rapid prototyping are applied to these functions, tremendous time, cost, and quality gains will result, while new, innovative capabilities emerge.

Personal Benefits

The listed benefits relate to those that a corporation would receive. However, rapid prototyping also offers benefits to the designer, mechanical engineer, or manufacturing engineer. Since rapid prototyping builds directly from an STL file, the process of prototyping is quick and convenient. This makes the power of the technology readily available, even when deadlines loom and time is short.

With increasing workloads, decreasing resources, and decreasing product development cycles, it is a challenge to maintain high-quality output. As the old saying goes, "Haste makes waste." Rapid prototyping helps people to do a great job even within a demanding work environment. The review of a physical representation of a sophisticated design offers the peace of mind of knowing that the work is good, error-free, and ready to be released to downstream processes.

CONCLUSION

This discussion of applications and benefits is only a meager representation of all that has—and will be—done with rapid prototyping and the value it delivers. Each person and each company has its own unique products, processes, goals, and challenges. From this individuality stems a wide range of applications and benefits.

While rapid prototyping can be, and has been, justified on the basis of the items discussed above, its ultimate power and reward arises when innovative applications address individual needs. Armed with this information, it is up to each reader to discover his or her own unique applications and benefits. However, as will be discussed in later chapters, the application of rapid prototyping must be balanced with a consideration of other technologies and processes to determine which is truly best for the task at hand.

CHAPTER 6

Rapid Prototyping versus CNC Machining

While this book is about rapid prototyping, it by no means suggests that this technology is the best or only solution for all applications. On the contrary, the best solution for each particular application can be determined when there is a thorough understanding of rapid prototyping and computer numerical control (CNC) machining.

As indicated earlier, many include the CNC machining in the collection of rapid prototyping technologies. In that machining can produce prototypes quickly, it is indeed rapid prototyping. However, the lack of distinction between rapid prototyping and CNC machining creates confusion in the marketplace. Since both technologies can be described as rapid prototyping, the discussion here adopts this naming convention: The acronym "RP" is used for additive rapid prototyping technologies, and "CNC" is used for subtractive machining technologies, as shown in *Figure 6-1*.

It was noted earlier that this book reviews only additive technologies because of the sheer volume of information that would be required to accurately depict, describe, and review both additive and subtractive technologies. With the addition of all machining processes and their respective capabilities, a thorough and complete comparison would fill a multi-volume reference collection. With this rationale, the decision to exclude CNC from earlier and subsequent chapters is not an indication that it is not rapid.

FIGURE 6-1. CNC machining of an aluminum workpiece with a 3-axis Makino Max 65 S®. *(Courtesy Leyshon Miller Industries)*

PRAGMATIC DECISIONS

When evaluating the needs of a prototyping project and selecting an appropriate method, it is common, but short-sighted, to use the technology with which there is past experience. Change is difficult for many, especially when the change is to a new technology that requires time and effort to learn and understand its strengths and limitations. Conversely, when a company adopts a new technology, often the use of the technology is seductive, and every project looks like a perfect fit.

In the early days of RP, it was common for the first few successes to open the floodgates of projects. Users would report extremely high demand for RP technology. Some applications would be perfect fits, while others were questionable. In extreme cases, the RP demand would be so high that it exceeded capacity and extended delivery times. Meanwhile, a 3-axis CNC would be idly waiting for the next job. The best solution is to apply the right technology. "Right" means reasonable time, reasonable cost, and acceptable results.

Selecting the right technology requires knowledge of, and experience with, both RP and CNC. Regrettably, many designers, engineers and manufacturing engineers do not have a thorough understanding of each technology. To begin to break down this barrier, the following information offers a head-to-head comparison.

PROCESS COMPARISON (Grimm and Wohlers 2003)

Rapid prototyping comes in many sizes, with differing prices that vary with the degree of capabilities. This is also true, to an even greater degree, for CNC machining where systems range from bench-top devices to huge gantry mills and from 2.5–5-axis high-speed machining centers, as illustrated in *Figures 6-2* and *6-3*. With this scope of technology, it is not appropriate to offer definitive comparisons. Therefore, the following information is offered on a relative basis. It is intended for a general comparison to help position the two classes of technology.

Output Quality

It does not matter if a technology is fast, inexpensive, or easy to use if the results do not meet the needs of the application. To understand if the output quality of RP and CNC are suitable, it is important to compare materials, part size, complexity, feature detail, accuracy, and surface finish.

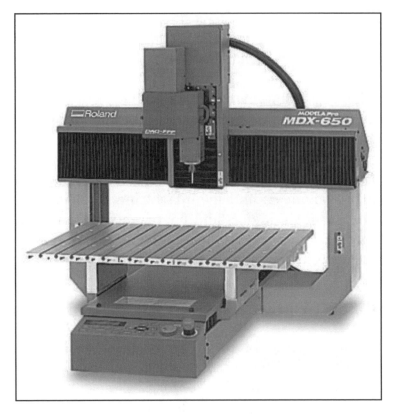

FIGURE *6-2.* CNC machines include benchtop systems such as the Roland MDX-650®, which offers 3-axis milling, and an optional rotary axis for 4-axis machining. *(Courtesy Roland DGA)*

When using this information, it is imperative to define "realistic" needs. Arbitrarily deciding that dimensional tolerance must be held to +0.001/–0.000 in. (+0.03/ –0.00 mm) or randomly choosing a production-grade resin artificially limits the technology options. If these are real requirements, then evaluate RP and CNC with respect to these parameters. If not, make the demands more realistic and expand the available options.

FIGURE 6-3. The Makino Max 65 S® is one example of a 3-axis vertical machining center. *(Courtesy Leyshon Miller Industries)*

Materials

RP systems are limited in the number of materials offered. Unlike CNC, each technology requires a unique material formulation for processing. This results in a limited scope of materials for each RP system. CNC, on the other hand, has much more latitude and, therefore, offers greater selection.

While the RP material advances have been significant and have led to use of materials such as plastics, metals, ceramics, and composites, any given system is likely to process just a single class of material. Within that class, the options may be limited to only a few alternatives. For example, stereolithography has more than a dozen material options, but each is limited to the photopolymer family. Fused deposition modeling, on the other hand, processes thermoplastics, but in only four different grades. Additionally, the material properties of an extruded, semi-molten material or sintered powder will not exactly match those of a thermoplastic molded under high pressure and high temperatures. At best, RP systems can closely approximate the material properties of production materials.

The CNC material selection is substantially larger than that of RP. With the right cutter, coolant, and feed and spindle speeds, nearly any plastic, metal, ceramic, or composite can be machined. This means CNC machining can process most production-grade resins and alloys. So, the available materials and material properties are nearly limitless. However, as with RP, subtractive methods will produce material properties that differ from those of a production casting or molding method. Yet, the difference in material properties will be significantly less than that for the additive processes for RP.

Maximum Part Size

Rapid prototyping has yet to match the breadth of part size offered by CNC. In the world of machining, the capacity of systems varies from inches to yards (millimeters to meters). For commercially available RP systems, the maximum part size ranges from 8 × 8 × 8 in. (203 × 203 × 203 mm) to 24 × 36 × 20 in. (610 × 914 × 508 mm). While this capacity may appear limiting for a car bumper, for example, the size range actually accommodates a majority of the typical applications for metal and plastic parts. From small desktop units to large gantry systems, CNC can fabricate parts and molds of virtually any size. Practically speaking, size is only limited by the capacity of the available machine tools and the availability of stock material in the required size.

Part Complexity

Due to the additive nature of RP, design complexity is rarely a consideration for prototyping. Production of even the most complex designs is possible, often with little impact on time or cost. With few exceptions, RP systems can produce any geometry that a CAD system can create. With CNC machining, complexity can limit the ability to produce the prototype while driving up time and cost. If design software can model a prototype, it is possible for RP to build it with little impact on time or cost. The ability to quickly and cost effectively produce complex parts is one of RP's biggest benefits. However, this benefit can become a liability if the prototype incorporates design elements that cannot be manufactured, as shown with the complex shape in *Figure 6-4*.

The CNC machining must deal with every feature in a part, and this can add time and cost. As the complexity of the part rises, so do the number of setups and tool changes. High aspect ratio features, deep slots and holes, and square inside corners can challenge the most expensive CNC system. While a 5-axis mill and some ingenuity can overcome challenging features, something as simple as an undercut can produce problems and lead to additional manufacturing steps.

Feature Detail

With the accuracy and precision of CNC systems, they generally outperform RP systems in the area of feature details. Yet, there are some exceptions.

In the area of sharp outside corners and smooth radii, CNC is more capable than RP. Yet, due to the additive nature of the technology, there are some features that RP can produce that cannot be matched with CNC machining. When applying a cutter to a block of material, all inside vertical corners will have a radius, whereas in RP, most

FIGURE 6-4. RP can produce complex shapes with little impact on time or cost. With CNC machining, this part would be impossible. *(Courtesy Z Corporation)*

systems can offer sharp inside corners. RP (axis dependent) can also produce small holes of unlimited depth, such as a 0.003-in. (0.08-mm) hole that is 3 in. (76 mm) deep. Finally, RP can produce small features with high aspect ratios. For example, RP is capable of producing a wall or rib with a thickness of 0.020 in. (0.51 mm) and a height of 2 in. (51 mm). In the CNC systems, this is often not achievable since the forces of cutting cause the wall to deflect during machining.

For all other features, CNC has the advantage over RP. Sharp edges, smooth blends, and clean chamfers are among the details where CNC excels. This is especially true when evaluating detail in terms of accuracy and surface finish.

Dimensional Accuracy

With CNC systems, extremely tight dimensional tolerances are possible and feasible. RP, on the other hand, typically delivers accuracy of 0.005 in. (0.13 mm) or worse. Additionally, RP tolerance is often a function of the size of the part. As part size increases, dimensional accuracy declines. It is common for RP vendors to claim accuracies of ±0.005 in. (±0.13 mm), and this is achievable in optimal situations where the size of the part is limited. With the variables that control the accuracy of RP production— system calibration, part size, age of material, temperature, humidity, and operator skill—it is unlikely that every prototype will adhere to this level of accuracy. Generally speaking, across all systems, a tolerance range of 0.005– 0.030 in. (0.13–0.76 mm) is common for RP systems.

With the right equipment, it is possible for the CNC systems to machine at high precision. While CNC in general is more accurate than RP, precision is commonly a function of the cost of the machine and the investment in the part that is produced. Yet, CNC systems will commonly and cost-effectively deliver dimensional accuracies in the ±0.001–0.005-in. (±0.03–0.13-mm) range.

Repeatability

Repeatability is a measure of the consistency of output. Given identical parameters, in a perfect world there would be no variation between two prototypes produced at different times. Without consistency of output, it is difficult to apply quality measurement and improvement techniques. While repeatability may not be of tremendous importance for a one-off prototype, as the prototype quantity increases and the product moves toward production, repeatability is vital.

Rapid prototyping is generally considered to have a low level of repeatability. If a given part is built on two different days or on two different machines, the results may vary. This occurs because of the dependency on many factors that directly impact the output of the technology. Changes in temperature or humidity can have dramatic effects on prototype quality. In RP, simple decisions regarding part orientation and location within the build envelope can also yield measurable impact on the output. With RP, building a prototype on two identical systems can produce differing results.

The CNC systems are more repeatable than RP devices. If the tool path, tool, and materials are unchanged, the output is highly consistent. While there will be small deviations across multiple parts, these differences will be of a magnitude that can be addressed with quality control processes such as statistical process control (SPC) and feedback/feedforward control systems.

Surface Finish

The layered nature of RP technology delivers a surface finish that does not match that of CNC machining. Additionally, the individual RP technologies—extrusion, sintering, curing, and binding—may have other deficiencies in the area of finish.

For RP, without benching the part, some, if not all, surfaces will be rough. While RP systems now offer layer thicknesses of 0.0005–0.0010 in. (0.013–0.025 mm), layer striations and stair stepping continue to impact surface finish, as shown in *Figure 6-5*. If desired, secondary operations can improve the smoothness to any level, but this can alter the dimensional accuracy of the part. Also, these operations can add time and cost to the project.

FIGURE 6-5. Surface finish, as shown on these RP models, is often a limitation of the technology.

Individual RP methods may also further degrade surface finish. For example, the powder-based systems of selective laser sintering and powder-binder printing will produce surface roughness from the binding and fusing of the powder particles. In fused deposition modeling, the extrusion process yields depressions between each pass of the extrusion tip. An additional consideration is that the quality of the surface finish in RP often varies between top, bottom, and sidewalls.

Unlike RP, CNC can produce surface finishes suitable for prototypes, patterns, and tooling. With the proper selection of cutters, speeds, and feeds, usable surfaces are most often delivered without any need for secondary operations. However, the desired surface finish may add additional time and cost to the machining operation. To eliminate tool marks and improve surface finish, a finishing cut would be required.

Operations

While the user of a prototype is concerned only with the quality of the output, technicians, operators, and manage-

149

ment must consider the operational strengths and deficiencies of the RP and the CNC machining.

Staffing

In larger RP operations, it is not uncommon for a single individual to do all of the preprocessing and machine operation for multiple systems. The ratio of man to machine can easily be 1:3 and as high as 1:6. For smaller operations, it is not uncommon for one individual to be responsible for preprocessing, part construction, and post-processing.

With the exception of secondary operations (benching), RP requires little labor. Within a few minutes, it is possible to prepare files for part production and start the build. During the build, there is little or no operator attendance, so the system operator is free to run other devices or tackle unrelated projects.

While computer-aided manufacturing (CAM) software applications have improved for CNC machining, in most cases, they have not eliminated the need for human intervention. For difficult or challenging projects, a CNC programmer may spend hours defining the tool paths required to machine a prototype. Once the tool path is generated, additional labor is required for machine setup and operation. Like RP, the CNC operations can be unattended, but this can vary from machine to machine.

While software tools and improved system controls have had some impact on the reduction of the skill level of the CNC programmer and machinist, the CNC systems continue to be staffed by skilled technicians. RP, on the other hand, has seen a dramatic decrease in the skill level required for the operators. In some cases, individual designers can construct their own rapid prototypes.

Rapid prototyping technologies are certainly not staffed with minimum-wage employees, but when compared to CNC machining, the demand for highly skilled talent is somewhat less. This is true in part because less overall labor is required. But the most significant factor in rapid prototyping is that system operation has progressed from a black art toward the ultimate goal of being a push-button science.

The CNC machining takes skill, creativity, and problem-solving abilities. From designing tool paths and machining strategies to operating and monitoring the machine tool, the work of experienced artisans with strong knowledge of materials and how they behave when cut is required.

Reliability

For most technologies, the reliability of systems improves over time. With years of experience and fine-tuning, all but the most sophisticated technologies can be made highly reliable. Take, for example, automobiles. Everyone knows that purchasing a car in its first model year carries some risk. The first-year vehicle is much more likely to have recalls and component failures since its field-testing is limited in scope. By the second year, many of the defects are designed out of the system.

Since the RP systems have, at most, 15 years in the field, and since new models are released frequently, they lack the time and history that allows weaknesses to be designed out. Additionally, many components in the system are designed specifically for the application or modified from others. This makes RP systems less reliable than other shop equipment. CNC, on the other hand, has decades of research and development behind it. Technologies in this class are mature and have the dependability and reliability that comes with time and with experience. Over the years, continuous improvement has eliminated system elements that diminish reliability.

Time

In general, RP is perceived to be very fast when constructing prototypes. However, this perception usually relates to just one component of the total process time, and does not take into account factors such as design complexity and physical size. In general, it is true that across all parts where prototypes are desirable, RP is the faster of the two technologies. But this is a dangerous assumption. Broad statements about the relative speed of RP and CNC machining are misleading, since many factors affect the most important aspect of time, total cycle time between request and delivery.

Construction Time

With many factors affecting part construction time, it is not advisable to state that one technology is faster than the other. Additionally, for each technology there are different factors that affect time. For example, the RP time is a function of size and part volume, while the CNC machining is a function of complexity and the amount of material to be removed.

Rapid prototyping system time is primarily a function of the amount of material to be added to the part, and in most cases the height of the part. For the CNC machining, time is primarily a function of the amount of material to be removed, the rate of material removal, and accessibility of the feature to be machined. When comparing the technologies, these factors lead to the generalization that for small parts (low material volume) with many features (complexity), RP is faster than the CNC machining. For example, if a 3-in. (76-mm) block with one through hole on each face is manufactured on both systems, the CNC machining would be significantly faster (it only needs three holes bored into a 3-in. [76-mm] block of material). If that same block were shelled out to a wall thickness of 0.060 in. (1.52 mm), the top face removed, and one boss added to each of the five remaining faces, RP would be the hands-down winner.

But there are other significant factors in construction time. RP has the advantage that it can construct multiple parts in a single operation and the total time is far less than the cumulative time of all the parts if built individually. The CNC machining is typically a sequential operation, one part at a time, where the cumulative time equals the time for each individual part. Another factor is that, for the most part, speeds, feeds, and acceleration do not impact RP systems. The technology creates the geometry at a rate that is basically fixed. For the CNC systems, cutter speed, feed rates, and tool acceleration impact the total cutting time. In CNC machining, the construction time for a part would vary if, in one instance, its shape allows long cuts where the tool can reach maximum velocity, but in another instance, the shape forces many directional changes in the tool path, which limits travel velocity.

A final consideration is the impact of setups. With RP, there is one setup for the entire operation, and it is often straightforward. The RP system is prepared (material fill, warm up, and calibration) and released. For CNC systems, there may be multiple setups for a prototype, and these setups take time.

Preprocessing Time

Both RP and the CNC machining require file processing to prepare for part construction. For RP, this process uses automated processing software to apply supports (where applicable), apply build parameters, and slice the data. In CNC machining, the processing phase uses CAM tools to define and generate tool paths.

As with the construction phase, preprocessing for RP is independent of part complexity. Rather, the automated processes time to completion is mostly a function of file size and computer speed. While experienced operators faced with a challenging prototype may invest time to manipulate the parameters for the automated processes, many, or most, prototypes can be processed with little user intervention, as illustrated in *Figure 6-6*. This allows the preprocessing phase to be completed in well under an hour. In addition, when processing multiple parts, the time is not cumulative. Beyond the operator's time to support and orient each part, the processing time will be relatively the same whether one or one hundred parts are preprocessed.

When generating tool paths for the CNC systems, an experienced machinist or toolmaker processes one file at a time. Experience weighs heavily, because the decisions made in tool-path generation directly affect time and quality. Even something as basic as cutter selection will affect the construction time and possibly the quality. Unlike RP systems, tool-path generation is a function of design

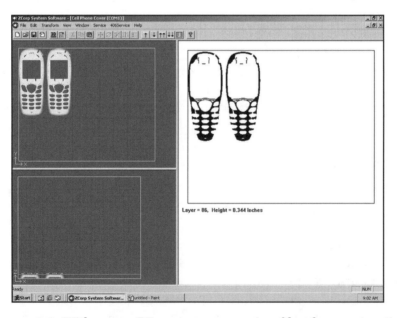

FIGURE 6-6. With many RP systems, preparing files for construction may be as simple as orientation and slicing. These cell phone housings were prepared in less than five minutes. *(Courtesy Fisher Design)*

complexity. The CAM programmer must decide the best way to machine the part, taking into account the number of setups, the number of tool changes, and the capabilities of the CNC device on which the tool path will be run. For every boss or rib added to the part, the programmer will spend more time in developing the tool path.

It must be noted that there are new software programs for the CNC systems that automate the tool-path generation process. For straightforward application, it is possible to let the software application do all the work. When this is possible, CNC preprocessing is a function of the same factor as RP, file size.

Post-processing Time

Post-processing includes all steps completed once a proto-type is removed from the machine. For RP, this often includes cleaning and part finishing. For the CNC machining, post-processing tends to be an optional step. For both technologies, the amount of finishing is prescribed by the intended application.

Without exception, RP systems demand some preparation of the prototype prior to delivery. For many systems, this includes removing residual material and supports. When an RP part requires smooth surfaces, the prototypes must be hand finished. This benching operation is labor intensive and often time consuming. For many applications, cleaning and finishing are measured in hours and sometimes in days.

The CNC machining does not demand the efforts and time for cleaning and finishing. In many cases, the prototype is only given a quick cleaning to remove debris. While high-end applications may demand additional finishing, this tends to be the exception. In general, the post-processing phase for CNC machining is measured in minutes.

Total Lead Time

The total lead time from data receipt to part delivery is defined mainly by the cumulative factors of construction time, preprocessing, and post-processing. Yet, there are other factors to consider.

With the exception of post-processing, RP is not labor intensive, and the majority of the work can be performed in an unattended mode. This allows RP systems to produce prototypes at all hours of the day and on working days and weekends without a second or third shift. This gives RP systems a tremendous advantage when compared to CNC operations that require staffing. Theoretically, RP systems could produce prototypes 8,760 hours a year. Realistically, many prototype shops plan for a maximum utilization of 7,000 hours per year. For the CNC systems that lack a fourth or fifth axis or automatic tool changers, a three-shift operation has a maximum load of 6,025 hours (excluding weekends and holidays).

While extremely dynamic and often challenging, the scheduling of RP systems can be much simpler than the CNC systems. For RP there are three primary steps: preprocessing, construction, and post-processing. For the CNC machining, the scheduling effort requires more coordination and consideration. For example, to begin part construction, the CNC systems require that operator, machine, materials, fixtures, tool path, and cutting tools are available. With the additional scheduling burden, it is much more likely that a work-in-progress queue results. When a job sits in a queue, the clock keeps ticking even though no work is performed. This extends the total cycle time of the CNC systems.

Cost

With the expense of purchasing, implementing, and operating an RP system, it would appear that it is much more expensive than the CNC machining. For some systems, this is certainly true. Yet, cost can be measured in many ways. It can be measured as total annual expense, as a cost per unit, or as a cost per hour. The method of measuring cost can change the determination of which technology is less expensive.

Although some RP systems may be more expensive than the CNC machines in initial capital expense and operating expense, the fully burdened hourly cost of RP can be less than that for the CNC machining. RP utilization is measured as a percent of all hours in the year, not work days or shift hours. With this single factor, RP can gain a threefold advantage over a CNC operation with a single shift. Thus, for an RP department operating at full capacity and high utilization, cost could be three times that of a CNC operation and yield an equivalent cost per hour.

When labor is factored in, the cost difference is further diminished. Since RP requires much less labor, especially from skilled operators, the reduction in labor expense further decreases the cost per part, cost per hour, and cost per year. In a shop where labor is 20% of the CNC machining cost, the combination of increased utilization and reduced labor can give RP as much as a fivefold advantage. In other words, RP could be five times more expensive in terms of purchase price, operating expense, and material cost, but still be less expensive than the CNC machining.

A final consideration in terms of cost is the expense of delays in product launch. For big-ticket items or high-volume consumer goods, a single day lost in the selling cycle may yield lost revenues that measure in the millions. With a typical gross profit of 40–50%, this could mean that a one-day delay in product launch could cost a company hundreds of thousands of dollars. In this scenario, the time advantage of the technology yields the greatest profit.

SUMMARY

With all of the exceptions and apples to oranges comparisons, there is no clear winner when choosing between RP and the CNC machining. At times, one will have an obvious advantage; at others the tables will turn. Thus, the conclusion is that the decision to use RP or the CNC machining is as unique and individualized as the products a company produces.

Perhaps the only true conclusion is that for each project, each product, and each department, the pros and cons of both RP and CNC machining should be evaluated. When both technologies have such great attributes, and they excel in diverse situations, it would be unwise at many companies to include only one technology in the design and

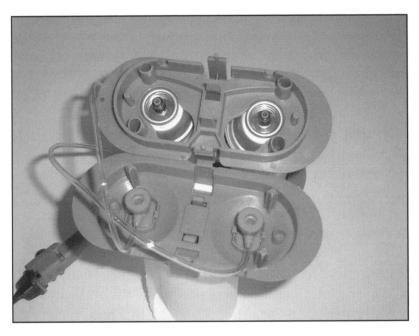

FIGURE 6-7. For many projects, the best option may be a combination of RP and the CNC machining, as shown in this functional prototype assembly. *(Courtesy Leyshon Miller Industries)*

manufacturing process. Most companies would be best served with ready access to both RP and CNC machining, producing prototypes that combine both technologies, such as the assembly illustrated in *Figure 6-7*.

REFERENCE

Grimm, Todd and Wohlers, Terry. 2003. "Is CNC Machining Really Better Than RP?" *Time-Compression Technologies*, January/February: 33–34.

CHAPTER 7

Rapid Prototyping Technologies
(Grimm 2002a; Grimm 2002b)

Choosing between the various rapid prototyping technologies can be difficult, and it has become even more challenging as the technologies have matured. Strengths and weaknesses are inherent in all processes, and rapid prototyping is no exception. To select the rapid prototyping method for a specific application requires a thorough understanding of the limitations as well as the strengths of each technology. The challenge is to select the right process for the task. Without hands-on experience, many of these factors will be obscure. Even with a full understanding of the parameters, there are trade-offs between the technologies that cloud the decision-making process.

The following review of four rapid prototyping technologies describes and compares the processes and offers guidelines that should be considered when choosing a technology. There are numerous issues to consider when selecting a process for a given application. To achieve the goal of prototyping quickly, accurately, and cost effectively, each aspect of the technology should be considered.

SYSTEMS

The four technologies for review and comparison have been selected on the basis of the number of installed systems. The four leading technologies, by total system sales, are:

- Stereolithography (3D Systems, Inc.)
- Selective laser sintering (3D Systems, Inc.)
- Fused deposition modeling (Stratasys, Inc.)
- 3D printing (Z Corporation)

For clarity purposes, the Z Corporation technology will be referenced as powder-binder printing. Unlike 3D Systems' and Stratasys' technologies, Z Corporation has not applied a unique name to its process. Instead, the company calls their process 3D printing. Since 3D printing is a generic term applied to many technologies from many vendors, this terminology could become confusing when comparing the technologies.

PROCESS OVERVIEW

Stereolithography

Developed by 3D Systems, Inc., stereolithography is the most widely used rapid prototyping technology. The company manufactures and markets a family of systems, including the Viper si2™, SLA® 5000 and SLA® 7000.

Stereolithography enables the creation of complex, three-dimensional models by successively "laser curing" cross-sections of a liquid resin. An ultraviolet (UV) laser contacts the resin, which is a photopolymer, causing the material to solidify. Although stereolithography is limited in its range of applicable materials, it is widely used for conceptual visualization, form and fit analysis, pattern creation, and light functional testing.

Selective Laser Sintering

Developed by the DTM Corporation, selective laser sintering (SLS®) is widely used for functional applications. DTM, now owned by 3D Systems, manufactures and markets the Vanguard si2™ system, and formerly Sintersta-tions™, for rapid prototyping and rapid tooling applications. The selective laser sintering process creates three-dimensional objects, layer by layer, from powdered materials. Heat from a CO_2 laser fuses (sinters) the powder within a precisely controlled process chamber. A major distinction between this and other rapid prototyping technologies is the wide variety of materials that can be utilized, including plastics and metals. Many of these materials prove suitable for functional analysis.

Fused Deposition Modeling

Developed and manufactured by Stratasys, Inc., fused deposition modeling (FDM®) is available in a number of systems. These include the FDM Maxum™, FDM Titan™, FDM 3000™, and Prodigy Plus™. Fused deposition modeling offers functional, plastic prototypes of acryloni-trile-butadiene-styrene (ABS), polycarbonate, and other materials. These thermoplastics are extruded as a semi-molten filament. This filament is deposited on a layer-by-layer basis to construct a prototype directly from 3D computer-aided design (CAD) data. The technology is commonly applied to form, fit, and function analysis, and concept visualization. Stratasys also offers a 3D printer under the Dimension™ brand.

Powder-binder Printing

Z Corporation's 3D printing technology (powder-binder printing) originated from research and development work at the Massachusetts Institute of Technology (MIT). From the original MIT license for its 3DP process, the Z Corporation devices, which include ZPrinter™ 310, Z®406, and Z®810, have taken the lead in terms of speed. Using drop-on-demand jetting technology, which is very similar to inkjet printing, the devices deposit a binder onto a powder bed. As the binder solidifies, a layer of solid geometry is created. Z Corporation also has taken the lead with the first devices to produce prototypes in color. The technology ranges from a low cost, small device for 3D printing to a larger device with greater capacity and throughput. With consideration of material properties and the speed of the technology, Z Corporation's systems are most often applied to concept models and form and fit prototypes.

PROCESS DETAIL

Stereolithography

The stereolithography process is illustrated in *Figure 7-1*. To prepare for a stereolithography build, liquid photo-polymer resin is added to the machine to bring the level up to operating condition. Upon launch, the build platform lowers into the vat of resin, leaving a gap equal to one layer thickness between the platform and the top surface of the resin. The system is now ready to construct the part's geometry.

A stationary UV laser, formerly gas charged and now solid state, is directed to X-Y scanning mirrors. The mirrors redirect the laser beam for travel in the X-Y plane. As the laser scans across the top surface of the resin, it imparts UV energy into the photopolymer, which causes the material to solidify. The depth of curing is a function of the laser power and the travel velocity. Longer laser dwells result in deeper cure depths. Typical layer thicknesses range from 0.001–0.010 in. (0.03–0.25 mm).

The first solidified layer is comprised entirely of support structures. This first layer of supports cures and forms a mechanical bond with the build platform. After constructing approximately 0.25 in. (6.4 mm) of support structure, the first layer of part geometry—the bottom of the part—is solidified. To create the part's geometry the laser

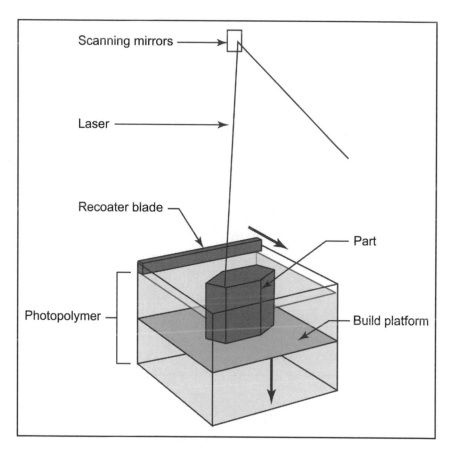

FIGURE 7-1. Diagram of the stereolithography process.

traces the boundaries of the profile and then solidifies the internal area with overlapping passes in the X and Y axes. After a layer is complete, the build platform lowers by one layer thickness, and the liquid resin flows over the top of the part. To level the resin, a blade sweeps over the surface. In preparation for curing of the next layer, sensors check the resin level and the laser power. The process repeats for subsequent layers until the part is complete. To remove the completed prototype, the platform is raised above the resin vat, and the platform, with the attached prototype, is removed from the build chamber.

Selective Laser Sintering

The selective laser sintering process is illustrated in *Figure 7-2*. To prepare for a selective laser sintering build, powdered material is added to the feed chamber. The build chamber is then brought to operating temperature, which is slightly below the melting point of the material.

Upon launching the build, the build piston lowers into the build chamber by one layer thickness. A feed piston rises to present powder, and a powder-delivery roller traverses the build chamber, depositing the powder into the gap between the build piston and the working surface. A stationary CO_2 laser, ranging from 50–100 mW, is directed to scanning mirrors that redirect the laser for travel in the X-Y plane. As the laser scans across the powder surface, it imparts thermal energy into the powder bed, which causes

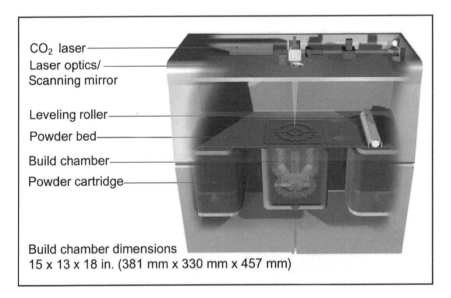

CO₂ laser
Laser optics/
Scanning mirror

Leveling roller
Powder bed
Build chamber
Powder cartridge

Build chamber dimensions
15 x 13 x 18 in. (381 mm x 330 mm x 457 mm)

FIGURE 7-2. Diagram of the selective laser sintering (SLS) process. *(Courtesy 3D Systems)*

the material to sinter (fuse). The depth to which the powder sinters is a function of the laser power and the velocity of travel. The longer the laser dwells, the deeper the sintering depth. Typical layer thicknesses range from 0.004–0.006 in. (0.10–0.15 mm).

For selective laser sintering, there is no requirement for support structures since the unsintered material remains in the build chamber and surrounds the part. Therefore, the first layer constructed is actual part geometry (the bottom of the part). After completing a layer, the build piston lowers, and the roller spreads a fresh layer of powder across the build chamber. The build process then repeats for all subsequent layers. Upon completion of the part, a cool-down cycle decreases the build chamber's temperature. To remove the part, the build piston rises and the part—with the surrounding powder cake—is removed.

Fused Deposition Modeling

The fused deposition modeling process is illustrated in *Figure 7-3*. This process does not require material replenishment prior to a build. However, if the material supply is insufficient to complete a build, or if the supply is exhausted prior to the end of a build, a new filament cartridge must be inserted into the device. The build chamber temperature is raised to operating level, which is just below the melting point of the material.

Unlike the other processes, parts are constructed in open space. The parts are not surrounded with liquid or powdered materials. Therefore, upon launching a build, the platform is not lowered by one layer thickness. Instead, its position provides a gap between it and the extrusion tip.

The building process uses an extrusion head, called the liquefier, which travels in the X and Y directions. Through this heated head passes a thin filament of build material. In a semi-molten state, the filament is extruded through the tip

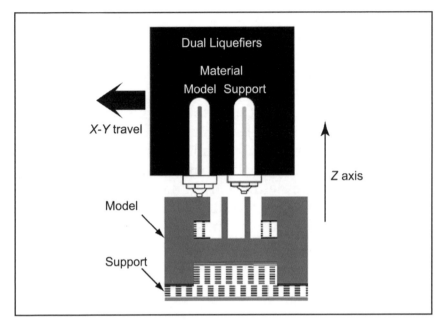

FIGURE 7-3. Diagram of the fused deposition modeling process. *(Courtesy Stratasys, Inc.)*

(nozzle). Each pass of the extrusion tip deposits a "road." The thickness and depth of the road are defined by the diameter of the extrusion tip and the velocity of the extrusion head. Upon deposition, the material bonds to the previous layer and rapidly hardens. As a layer is completed, the build plate lowers by one layer thickness, typically 0.005–0.012 in. (0.13–0.30 mm).

Unlike stereolithography and selective laser sintering, the FDM process is nearly continuous. The extrusion process proceeds from one layer to the next with little or no delay.

Like stereolithography, the process requires support structures. These are the first elements to be deposited. They are fixed to a disposable mat placed on the build platform. Starting with the bottom of the part, the first layer of geometry is deposited, and the process is repeated for all subsequent layers. When finished with the build, the part, fixed to the disposable mat, is removed from the build chamber.

Powder-binder Printing

The powder-binder printing process is illustrated in *Figure 7-4*. To begin the process, the binder supply is replenished and fresh powder is added to the feed box. When a build is launched, the build platform lowers into the build chamber

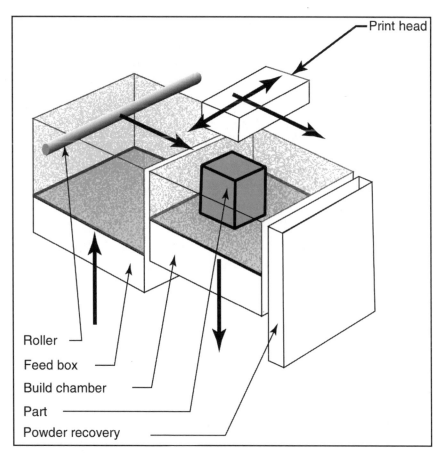

FIGURE 7-4. Diagram of the powder-binder printing process as developed by Z Corporation.

to a depth of one layer thickness, typically 0.0035–0.0070 in. (0.089–0.178 mm). A piston rises in the feed chamber to present fresh powder, and a roller spreads the material across the build chamber.

Like selective laser sintering, the process does not require support structures. Thus, the build begins with the first layer of the part. The process uses an inkjet print-head assembly that deposits liquid binder. Traversing the build chamber on a gantry, the print head passes over the powder, depositing fine droplets of binder onto the powder bed. After each print head pass, the gantry moves across the build chamber for the next pass.

Unlike the other processes, powder-binder printing is a raster operation. With each pass of the print head, a curtain of binder droplets covers 0.50–2.00 in. (12.7–51.0 mm) of the layer profile. The print head can completely cover the build chamber of the Z406 in less than eight seconds and that of the the ZPrinter 310 in 30 seconds. While the size of the layer profile may increase time, since there are multiple print-head passes, the time per layer is relatively constant when compared to the other processes. With the speed of the printing operation, the most significant component of time is often that for spreading the powder.

For some devices, multi-color parts are possible. This is achieved with the deposition of binders dyed cyan, magenta, and yellow. Similar to the process used for 2D color inkjet printers, the combination of the three colored binders offers a palette of more than a million colors.

After completion of a layer, the build platform lowers and fresh powder is spread. The process is then repeated for all subsequent layers. After the last layer is solidified, the part rests in the build chamber to allow the binder to harden and the part to dry. Then the surrounding powder is removed and the part is extracted.

COMPARISON OF PROTOTYPE PROPERTIES

When evaluating rapid prototyping processes, the first consideration is whether the technology can provide a prototype that satisfies the physical requirements of the project. Physical considerations relate to the quality of the prototype and its ability to match performance demands.

Material Properties

When asked to rank areas of importance, rapid prototyping users often state that material properties are the most important factor. To address the needs of industry, material properties matching those of the intended production material, or suitable for the application, are important. While great strides have been made since the early days when rapid prototypes were very brittle, there continues to be room for improvement, and material vendors have responded with a significant amount of research and development.

If material properties are a critical parameter for success, it is best to select the appropriate materials and then evaluate the capabilities of the technologies that utilize these materials. Additionally, many system parameters, such as build time and dimensional accuracy, are material dependent. Therefore, it is recommended that the systems be compared with consideration of the device and material to be processed.

It is important to be cautious when reviewing vendor-supplied material properties. Although testing adheres to standards, variances in build parameters, machine type, and elapsed time can yield significant deviations in the results. For example, a supplier of rapid prototyping materials reports that a variation in hatch overcure, a user-specified parameter, can alter the tensile modulus of stereolithography materials by 30%. With this latitude, vendors may select the testing conditions that present their materials' most favorable results.

Stereolithography

The stereolithography process is limited to photopolymer materials, which are solidified with UV light. Unlike the other processes, stereolithography lacks a breadth of material classes. Yet, the process has the largest number of available materials. Stereolithography materials are available from an open market where multiple companies, including 3D Systems, develop materials. Currently, there are more than 24 materials available from the three leading material suppliers.

There have been significant developments in the materials available for stereolithography. The advancements have improved physical properties and operating parameters, and have extended the list of appropriate applications. Once known as the technology that produced brittle prototypes, stereolithography now offers a range of material properties well suited for form and fit evaluation and, in some cases, functional testing.

Stereolithography materials offer properties that range from flexible and durable to rigid and strong. While the properties do not match those of injection-molded materials, in many cases, the materials approximate the characteristics of thermoplastics like polypropylene, polyethylene, nylon, and ABS.

A unique feature of some materials is that they produce a clear prototype. An example is shown in *Figure 7-5*. The transparent prototypes are used for visual analysis of working features internal to an assembly. They are also used to visualize fluid flow in a variety of applications.

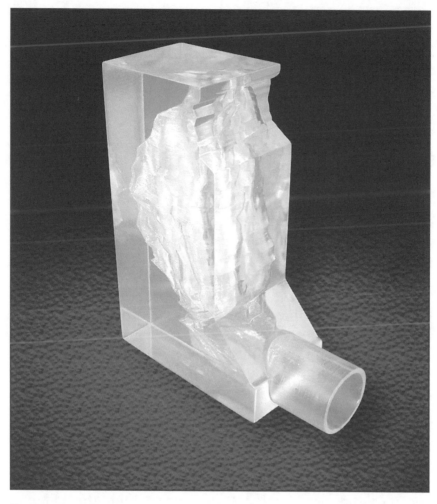

FIGURE 7-5. Clear prototypes are available from some rapid prototyping systems, as illustrated by this human nasal cavity produced by stereo-lithography. *(Courtesy Accelerated Technologies)*

Selective Laser Sintering

As with stereolithography, there have been advances in material technology for selective laser sintering. Overall, SLS offers the widest latitude in material properties, including plastics and metals. With the breadth of material properties that the technology delivers, SLS is often applied to functional testing and tooling applications. Unlike stereolithography, these materials are available only through 3D Systems.

For plastic prototypes, two polyamide-based materials are offered: DuraForm™ PA and DuraForm™ GF. The polyamides produce prototypes with nylon-like characteristics. While both share the same base material, DuraForm GF contains glass beads for additional strength and rigidity. Approaching the mechanical properties of injection-molded thermoplastics, DuraForm offers exceptional tensile, flexural, and impact strength. It is a suitable material for functional applications that require strength and rigidity.

LaserForm™ ST-100 is a binder-coated 420 stainless steel. During the selective laser sintering process, the binder is sintered. This green part is heated in a furnace to burn off the binder and infiltrate with bronze. The resulting part is approximately 60% stainless steel and 40% bronze. Applications for LaserForm include metal prototypes and tooling inserts.

Selective laser sintering also offers two application-specific materials: Somos® 201 and CastForm™ PS. Somos 201 is an elastomer used for flexible, rubber-like prototypes. CastForm PS, a proprietary polystyrene-based material, is used in investment casting applications. The CastForm part becomes the sacrificial pattern from which the ceramic mold is created.

Fused Deposition Modeling

As with selective laser sintering, material properties are one of fused deposition modeling's greatest strengths. While Stratasys manufactures all of the materials for the FDM process, each is produced from a custom blend of commercially available thermoplastic resins. All fused deposition modeling systems offer ABS as a material option. Nearly 90% of all prototypes produced by FDM are of this material. Users report that ABS prototypes demonstrate between 60–80% of the strength of injection-molded ABS. Other properties, such as thermal and chemical resistance, also approach or equal those of injection-molded parts. This makes ABS a widely used material for functional applications.

Use of a new material, polycarbonate, is growing rapidly. The additional strength of polycarbonate produces a prototype that can withstand greater forces and loads than the ABS material. Some users believe that polycarbonate produces a prototype that demonstrates the strength characteristics of injection-molded ABS.

There are other specialty materials for fused deposition modeling. These include polyphenylsulfone, elastomer, and wax. Polyphenylsulfone offers high heat and chemical resistance with strength and rigidity. Elastomer is intended for functional prototypes that behave like a "rubber" part with a durometer rating in the mid-to-upper range of the Shore A scale. Wax material is specifically for the creation of investment casting patterns. The properties of the wax allow the fused deposition modeling pattern to be processed like the traditional wax patterns used in foundries.

Powder-binder Printing

Like selective laser sintering, powder-binder printing uses powdered materials. However, unlike selective laser sintering, the materials are not powdered metals or plastics. Current offerings include two cellulose (starch) materials (zp14™ and zp15E™) and a plaster material (zp102™). Without further processing, these materials lack the desired mechanical properties for advanced applications. Yet, for the primary application of concept modeling, many find the resulting prototypes to be more than suitable.

After part construction, the prototype may be infiltrated to improve its physical properties and make it more suitable for applications that include form, fit, and light functional testing. The infiltration materials include epoxy resin, cyanoacrylate, urethane, and wax. The combinations of powder and infiltrant offer a broad range of material properties. For example, the zp15E material, when combined with the Synair Por-A-Mold® elastomeric urethane infiltrant, delivers a flexible, rubber-like prototype.

Color

Limited to a few technologies, color prototypes are a fairly recent development. For the most part, systems that offer this feature will produce the prototype in a single color. Of the systems reviewed, only powder-binder printing allows the production of prototypes with multiple colors selectively applied.

While color is an often-cited request, its application has been limited, perhaps because no system can deliver a color prototype that looks exactly like the end-use item. One interesting application of color is its use in finite element or mold-flow analysis. Rather than relying on a flat image with color bands indicating stress or flow characteristics, the prototype becomes a physical, 3D color plot of the analysis results.

Stereolithography

Most materials for stereolithography are amber or off-white. However, in recent years, clear and tinted materials have become available.

One manufacturer produces a resin that changes color when higher UV energy is imparted into localized areas of the prototype. With the additional UV energy, the natural color of this material turns to a distinctive red. The material's market is mostly the biomedical industry where it is used to create models that exhibit a growth, defect, or tumor highlighted with color.

Selective Laser Sintering

The base color of the raw material becomes the color of the selective laser sintering prototype. For the polyamides and Somos 201, this yields a white or slightly gray part. With the metal material, the prototypes can have a mottled appearance that blends the color of stainless steel and bronze. However, polishing the prototype produces consistent stainless-steel coloration.

Fused Deposition Modeling

Fused deposition modeling materials offer a wide variety of colors. However, each prototype is limited to a single color, which is determined by the base material. Including white, which is most frequently used, ABS comes in eight material colors. The color options include blue, yellow, orange, red, green, black, gray, and white. Translucent prototypes in clear, red, or yellow are also available. *Figure 7-6* shows an assembly of parts made from colored materials.

Powder-binder Printing

Powder-binding printing offers the greatest latitude in colored prototypes. With the addition of food-grade dyes to the binder, a wide array of colors is possible. Using a blend of cyan (blue), magenta (red), and yellow, powder-binder printing has a 24-bit palette that yields millions of colors. Unlike the other technologies, a single prototype can be produced in a multitude of colors. For example, the turbine blade shown in *Figure 7-7* is a multi-colored model depicting the results of finite element analysis (FEA).

Since the dye is added to the water-based binder, and since the binder does not fully saturate the powder, colors

FIGURE 7-6. Color can be advantageous for concept models, presentations, and even form, fit and function reviews, as shown in this tricycle
produced with fused deposition modeling. *(Courtesy Stratasys, Inc.)*

will lack vibrancy. The white color of the raw powder
subdues the printed color. This is most obvious when
printing a black part, where the resulting color is a dark
gray. However, when infiltrating a prototype with a clear
material, the color may become more vibrant.

Comparative Results

For direct comparison of several properties, a simple test
part was built in each technology (see *Figure 7-8*). The
results from these builds are discussed later in this chapter.
The data was gathered from part construction on: the Viper

FIGURE 7-7. Full color prototypes can describe the results of FEA analysis, as demonstrated by this turbine blade produced on a Z®406 system. *(Courtesy Z Corporation)*

si2 and SLA 7000 (stereolithography) using 0.006-in. (0.15-mm) layers with DSM Somos Watershed™ resin; the Vanguard si2 (selective laser sintering) using 0.006-in. (0.15-mm) layers with the DuraForm PA material; the FDM Titan (fused deposition modeling) with 0.010-in. (0.25-mm)

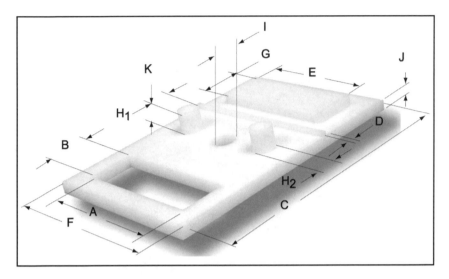

FIGURE 7-8. Diagram of the test part used for accuracy and time studies. The part measures 6.00 x 4.00 x 0.75 in. (152.4 x 101.6 x 19.1 mm).

layers using ABS material; and the Z406 (powder-binder printing) with layers of 0.004 in. (0.10 mm) using zp102 plaster material.

Dimensional Accuracy

The dimensional accuracy of a rapid prototype depends upon many factors, the most obvious of which are operator capability and system configuration. Other considerations include the time frame in which measurements are taken, environmental exposure, and the amount of post-process finishing work. Tolerance deviation also depends upon the axis along which measurements are taken.

Dimensional accuracy results for a test part are presented in *Figure 7-9* and *Table 7-1*. With all of the potential variables, user results will vary. Therefore, the information presented is a generalization of the dimensional accuracy that is reasonable to expect. An additional consideration is that the testing was performed on a single part, which is not

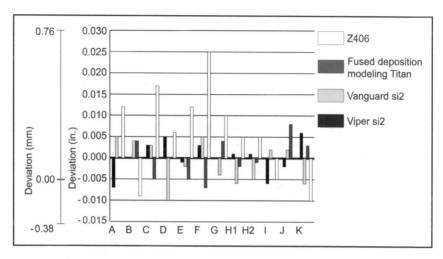

FIGURE 7-9. Accuracy data for the test part illustrated in *Figure 7-8* (Grimm 2002b).

a statistically sound sampling of parts. The goal of this comparison is to present dimensional accuracies a user may realize in a one-time, one-off mode of prototyping.

Stereolithography

Upon completion of the prototype and before post-processing, stereolithography tends to offer the highest degree of dimensional accuracy, matched only by fused deposition modeling. The shrinkage of the stereolithography epoxy resins is significantly less than that of the selective laser sintering and fused deposition modeling plastic materials. The stereolithography epoxy materials experience less than 0.1% shrinkage during the build process, making compensation simple to predict and easy to control.

Selective Laser Sintering

Selective laser sintering has reasonable accuracy, but results may vary with part geometry, size, and operational conditions. The materials have shrinkages of 3.0–4.0%.

TABLE 7-1. Dimensional accuracy data for each of the processes producing the test part in Figure 7-8 (Grimm 2002b).

	Nominal	Viper si2			Vanguard si2			FDM Titan			Z406		
		Actual	Deviation	%	Actual	Deviation	%	Actual	Deviation	%	Actual	Deviation	%
A	3.000 (76.20)	2.993 (76.02)	-0.007 (-0.18)	0.2	3.005 (76.33)	0.005 (0.13)	0.2	3.000 (76.20)	0.000 (0.00)	0.0	3.012 (76.50)	0.012 (0.30)	0.4
B	1.000 (25.40)	1.000 (25.40)	0.000 (0.00)	0.0	1.004 (25.50)	0.004 (0.10)	0.4	1.004 (25.50)	0.004 (0.10)	0.4	0.991 (25.17)	-0.009 (-0.23)	0.9
C	6.000 (152.40)	6.003 (152.48)	0.003 (0.08)	0.1	6.003 (152.48)	0.003 (0.08)	0.1	5.995 (152.27)	-0.005 (-0.13)	0.1	6.017 (152.83)	0.017 (0.43)	0.3
D	0.100 (2.54)	0.105 (2.67)	0.005 (0.13)	5.0	0.090 (2.29)	-0.009 (-0.23)	9.0	0.100 (2.54)	0.000 (0.00)	0.0	0.106 (2.69)	0.006 (0.15)	6.0
E	3.000 (76.20)	2.999 (76.17)	-0.001 (-0.03)	0.0	2.999 (76.17)	-0.001 (-0.03)	0.0	2.995 (76.07)	-0.005 (-0.13)	0.2	3.012 (76.50)	0.012 (0.30)	0.4
F	4.000 (101.60)	4.003 (101.68)	0.003 (0.08)	0.1	4.005 (101.73)	0.005 (0.13)	0.1	3.993 (101.42)	-0.007 (-0.18)	0.2	4.025 (102.24)	0.025 (0.64)	0.6
G	1.000 (25.40)	1.000 (25.40)	0.000 (0.00)	0.0	0.996 (25.30)	-0.004 (-0.10)	0.4	1.004 (25.50)	0.004 (0.10)	0.4	1.010 (25.65)	0.010 (0.25)	1.0
H1	0.500 (12.70)	0.501 (12.73)	0.001 (0.03)	0.2	0.495 (12.57)	-0.005 (-0.13)	1.0	0.498 (12.65)	-0.002 (-0.05)	0.4	0.505 (12.83)	0.005 (0.13)	1.0
H2	0.500 (12.70)	0.501 (12.73)	0.001 (0.03)	0.2	0.495 (12.57)	-0.005 (-0.13)	1.0	0.499 (12.67)	-0.001 (-0.03)	0.2	0.505 (12.83)	0.005 (0.13)	1.0
I	0.500 (12.70)	0.494 (12.55)	-0.006 (-0.15)	1.2	0.502 (12.75)	0.002 (0.05)	0.4	0.500 (12.70)	0.000 (0.00)	0.0	0.485 (12.32)	-0.015 (-0.38)	3.0
J	0.250 (6.35)	0.248 (6.30)	-0.002 (-0.05)	0.8	0.252 (6.40)	0.002 (0.05)	0.8	0.258 (6.55)	0.008 (0.20)	3.2	0.250 (6.35)	0.000 (0.00)	0.0
K	0.500 (12.70)	0.506 (12.85)	0.006 (0.15)	1.2	0.494 (12.55)	-0.006 (-0.15)	1.2	0.503 (12.78)	0.003 (0.08)	0.6	0.490 (12.45)	-0.010 (-0.25)	2.0
	Minimum			0.0			0.0			0.0			0.0
	Maximum			5.0			9.0			3.2			6.0
	Average			0.8			1.2			0.5			1.4

All dimensions in in. (mm)

The larger shrink rates increase the tendency for the prototype to warp, bow, or curl. Also, the selective laser sintering process is less predictable and controllable since it relies on raising the temperature of the powders to just below their melting points. Reliance on heat and heat transfer makes selective laser sintering dependent on chamber temperature, laser output, and heat retention within the previously sintered powder. Should elevated temperatures be present in the unsintered powder, sintering of the part may cause undesired materials to fuse to the surface.

Fused Deposition Modeling

In general, fused deposition modeling provides accuracies equal to those of stereolithography and better than those of selective laser sintering and powder-binder printing. The accuracy of fused deposition modeling is affected by fewer user-controlled variables. Compensation for parameters that impact accuracy is provided in the system's control software. Although the shrinkage rate of the materials equals or exceeds that of selective laser sintering, fused deposition modeling technology accurately predicts and compensates for this variable.

Powder-binder Printing

Dimensional accuracy is not a strength of powder-binder printing. While the plaster material is generally more accurate than the starch, both typically result in dimensional deviations higher than those of the other three technologies. Factors influencing dimensional accuracy include dispersion of the liquid binder, rate of absorption of the binder into the dry powder, and the compaction (density) of the powder in the build chamber.

For concept modeling, many find the accuracy reasonable. When applying rapid prototyping to a number of applications, some users will build powder-binder printing prototypes for early concept models and then use one of the other technologies for more demanding applications. Additionally, as with all of the rapid prototyping technologies, secondary finishing or machining may be used to improve dimensional accuracy for more demanding applications.

Stability

Stereolithography

Unlike selective laser sintering and fused deposition modeling, stereolithography parts are susceptible to additional shrinkage and distortion after part construction. Therefore, part measurements taken one week after production may vary from those taken immediately after the part was completed. Heat, moisture, and chemical agents can also affect photopolymer materials. Although stereolithography produces an accurate part, exposure to any one of these elements can negatively impact tolerances.

Selective Laser Sintering

The plastic prototypes produced in selective laser sintering are dimensionally stable once they are removed from the system and cooled. As with any plastic material, as long as excessive heat is avoided, the part will not shrink, warp, or distort.

When constructed of the metal material, parts are dimensionally stable once fully infiltrated. However, during the furnace cycle, where binder is burned off and bronze is infiltrated, dimensional accuracy can be degraded if care and caution are not exercised.

Fused Deposition Modeling

The properties of the fused deposition modeling materials do not change with time or environmental exposure. Just like their injection-molded counterparts, these materials retain their dimensional and physical properties in nearly any environment.

Powder-binder Printing

Without infiltration, starch and plaster prototypes are susceptible to changes in material properties and dimensional accuracy. Heat may dry the prototypes, making them brittle. Moisture in the air may be absorbed, causing the prototypes to soften, swell, or distort.

When infiltrated and completely sealed, the prototypes take on the stability characteristics of the infiltration material. In most cases, this makes the prototype stable in many environments and operating conditions. However, during the infiltration process care must be taken, since the addition of the infiltrant may result in part distortion.

Surface Finish

Surface finish is a critical consideration when applying rapid prototyping to a project. While all systems can produce a surface finish suitable for concept models, their finishes are not suitable for painted prototypes or tooling patterns. To achieve the desired finish for these applications, finishing and benching must be performed. The key difference between systems is how much labor must be invested to achieve the desired results.

Stereolithography

Of the four systems, stereolithography delivers the best surface finish prior to any benching operations. However, finish characteristics vary with surface orientation. Users of stereolithography find that the smoothest surface finish is on the top face of the prototype. The uppermost faces are contrasted by the rougher sidewalls and bottom surfaces.

The sidewalls of stereolithography parts exhibit striations between build layers. The bottom surface of a stereolithography part is affected by the support structures. When removed prior to a finishing operation, the supports leave rough areas and pits on the bottom face. The precision of stereolithography that originates from the controlled and predictable nature of the process, combined with good surface finish, will allow obvious detection of stair stepping, if it is not eliminated with part benching.

Selective Laser Sintering

Due to the sintering operation of the selective laser sintering process, all surfaces demonstrate rough and porous qualities. Since sintering fuses the powdered material without melting, voids are created between particles. Additionally, excess heat generated during the sintering process can cause further surface finish degradation. This heat results in undesired material sticking to the part surface.

Surface finish is also a function of the powder particle size. With finer powders, surface finish can be improved.

Fused Deposition Modeling

The most obvious limitation of fused deposition modeling is surface finish. Due to the extrusion of a semi-molten plastic, prototype surfaces exhibit a rougher finish than with stereolithography. The finish is comparable to that of the selective laser sintering and powder-binder printing processes. While improved surface finish is possible with smaller extrusion diameters and thinner layers, the top, bottom, and sidewalls will still show the contours of the passes of the extrusion tip and the build layers.

Users find it wise to consider surface finish requirements when orienting a part for building. Those surfaces that demand a higher level of finish are often oriented vertically.

Surfaces of lesser importance are oriented horizontally such that they are either a bottom or top surface.

Powder-binder Printing

All surfaces of a powder-binder printing prototype have a rough, textured appearance. The plaster material offers improvement in surface characteristics over the starch material, but the surface finish is still relatively rough. Users have characterized the finish of a plaster prototype to be that of 150 grit sandpaper. *Figure 7-10* illustrates the surface finish of a prototype after infiltration but prior to sanding.

Feature Definition

Feature definition is an important aspect of any prototype, whether applying it to concept modeling or functional analysis. When comparing feature definition, it is important to understand that there may be two states, the one which can be built and the one which is usable. The combination of support structures, part-finishing requirements, and material properties often dictates that the minimum feature size producible in the machine is smaller than that which will survive part benching and routine handling.

Stereolithography

In general, stereolithography is capable of producing a 0.010-in. (0.25-mm) feature, which is defined by process controls and the spot size of the laser. For most stereolithography systems, the laser has a spot size of 0.010 in. (0.25 mm). The Viper si2 is an exception. It was designed specifically for fine detail work. It has a minimum spot size of 0.003 in. (0.76 mm) and can capture finer details than the other 3D Systems devices. Although the system is capable of replicating micro features, removing the support material

FIGURE 7-10. Powder-based processes often have a rough, textured surface finish.

can be problematic. In some instances, small features will be damaged when supports are removed. For those that do survive, there is the possibility that a thin wall or tiny standoff may not withstand routine handling. Considering material properties, it is common to find that prototypes use a minimum feature size of 0.020 in. (0.51 mm).

Selective Laser Sintering

For selective laser sintering devices, as with stereolithography, the spot size of the laser dictates the minimum feature size. The laser has a diameter of 0.018 in. (0.46 mm). However, due to the nature of the technology, surrounding material often fuses to the part even if the laser does not directly contact it. This yields a typical minimum feature size of 0.025 in. (0.64 mm). Unlike stereolithography, this becomes the minimum usable feature size, since support structures are not a concern and the base material has reasonable properties.

In terms of metal parts, the minimum feature size does behave like stereolithography. From the machine and prior to infiltration, the green part has little strength. Therefore, evacuating powder that surrounds small features can easily result in damage. In general, a minimum feature size for a metal part or tool ranges from 0.030–0.100 in. (0.76–2.54 mm), where the size is a function of the aspect ratio of the feature.

Fused Deposition Modeling

Although advanced operators can produce smaller features, most prototypes are constrained to a minimum feature size that is twice the road width. Without user intervention, the process uses a "closed path" that limits the minimum feature size to two passes of the extrusion tip. For common tip sizes and build parameters, the minimum feature size ranges from 0.016–0.024 in. (0.41–0.61 mm). While larger than that of stereolithography, this range is in line with the usable minimum feature size of stereolithography. The fused deposition modeling minimum feature size is equal to or better than that of selective laser sintering. With material properties similar to injection-molded ABS or polycarbonate, fused deposition modeling can deliver a functional feature size in the 0.016–0.024 in. (0.41–0.61 mm) range.

Powder-binder Printing

With a print resolution of up to 600 dpi, the powder-binder printing process can deposit binder to create features in the 0.004–0.006 in. (0.10–0.15 mm) range. However, prototype features cannot be reproduced to this level. Two significant factors are the controlled spread of the binder and the strength of the powder/binder combination. Predicting and controlling the dispersion of the binder in the micro range is difficult due to variables such as powder density and powder moisture levels. In addition, liquids follow the path of least resistance, so the binder will disperse beyond the bounds of the geometry. Due to the fragile nature of the prototypes before infiltration, small features are difficult to achieve when powder removal is considered. In general, the plaster material can accommodate features of 0.030–0.060 in. (0.76–1.52 mm).

Machinability

Stereolithography

With photopolymer material advances, stereolithography parts now offer reasonable machinability. However, when milling, drilling, or tapping these parts, cutting tool feed and speed rates may require adjustment to prevent chipping or breakage.

Selective Laser Sintering

Selective laser sintering prototypes and tools produced in the thermoplastic and metal materials can be easily machined. However, with the polyamide-based DuraForm, the machined area can melt if cutting speeds are too high.

Fused Deposition Modeling

With little additional consideration, fused deposition modeling prototypes can be machined. To offset surface deficiencies and improve feature detail, users often perform a secondary machining operation to refine the details of the prototype when exceptional quality is necessary.

Powder-binder Printing

Without infiltration, powder-binding printing prototypes are too fragile for machining operations. However, once infiltrated, the parts assume many of the mechanical properties of the infiltrant. Therefore, with the proper selection of an infiltrant, prototypes may be machined.

Environmental Resistance

It is obvious that functional prototypes must withstand any environmental conditions to which the end item will be exposed. What is often overlooked is that the environment can also affect concept and form and fit models. Environmental considerations include temperature, moisture, and chemical agents.

Stereolithography

Photopolymers offer poor resistance to certain environmental conditions. For prototyping requirements, the most frequent requirements are resistance to temperature, moisture, and chemicals. In general, parts constructed in photopolymer resins should not be subjected to moisture, heat over 115° F (46° C), or many chemicals. Although new materials have drastically reduced sensitivity to moisture, some may demonstrate swelling and changes in the material properties when exposed to high humidity or immersion in water. Exposure to elevated temperatures will cause the parts to soften, which allows warping and distortion. Furthermore, most chemicals will deteriorate photopolymers.

It should be noted that the previous statements apply to the general-purpose resins available for stereolithography. New developments in resin technology have resulted in materials with high heat deflection temperatures, improved mechanical properties and, in some cases, greatly improved chemical resistance.

Selective Laser Sintering

Selective laser sintering prototypes provide material properties similar to those of the thermoplastics on which they are based, including resistance to environmental exposure. The polyamide materials can withstand moisture, heat up to 325° F (163° C), and many chemicals. With a surface sealant, these materials have been used in watertight and underwater applications without any indication of swelling due to absorption. Many selective laser sintering materials are reported to withstand exposure to chemical agents such as acids, bases, alcohol, hydrocarbons, ethers, and ketones.

Fused Deposition Modeling

Like selective laser sintering, fused deposition modeling prototypes offer material properties similar to those of the thermoplastics, on which they are based. This includes environmental and chemical exposure resistance. With the ABS material, users can subject prototypes to temperatures of 200° F (93° C) and chemical agents that include oil, gas, and even some acids. Additionally, on the Titan, polyphenolsulfone (PPSF) is a material option that offers heat resistance to 400° F (204° C) and greatly improved chemical resistance. Fused deposition modeling prototypes, like those from selective laser sintering, are unaffected by moisture, so they retain their original mechanical properties and dimensional accuracy.

Powder-binder Printing

In an uninfiltrated state, powder-binder printing parts should be limited to a controlled environment. Yet, even in such an environment, the prototypes are susceptible to moisture absorption that can degrade the prototypes qualities. When a prototype is infiltrated and fully sealed, it may be subjected to environmental conditions and chemical agents that are appropriate for the infiltrant.

Physical Size

Rapid prototyping systems offer a wide range of build envelopes, from less than 1.00 × 1.00 × 1.00 ft (0.3 × 0.3 × 0.3 m) to nearly 3.00 ft (0.9 m) per axis. While these systems may not be capable of producing a full-size car bumper, they do accommodate the majority of parts molded or cast for consumer and industrial products.

For parts that exceed the build envelope, it is common to section the STL file and build the prototype in pieces. After construction, the sections are bonded to create the finished prototype.

It is important to note that in some cases the advertised size of the build envelope is greater than the usable volume. While the platen and construction area are listed accurately, operating parameters may limit the build envelope to a smaller usable size.

Stereolithography

Stereolithography devices range from 10 × 10 × 10 in. (254 × 254 × 254 mm) for the SLA 250 and Viper si2 to 20 × 20 × 24 in. (508 × 508 × 610 mm) for the SLA 5000 and SLA

7000. While experienced users may take advantage of the entire envelope, most allow approximately a 0.5-in. (13-mm) border between the edge of the build platform and the prototypes.

Selective Laser Sintering

The Vanguard si2 and its predecessor, the Sinterstation 2500 plus, provide a usable build envelope of 13 × 11 × 15 in. (330 × 279 × 381 mm). The physical work envelope is larger than these specifications, but the usable work volume is limited by system operating parameters.

Selective laser sintering prototypes tend to be better suited for bonding than stereolithography parts. Due to the porosity of the part's surface, the adhesive is able to penetrate to form a stronger bond. Additionally, the plastic parts can be ultrasonically welded to produce a strong bond suitable for functional applications. It is also possible to weld the metal prototypes or tools.

Fused Deposition Modeling

Within its family of systems, fused deposition modeling offers a wide range of build envelopes where the advertised size is equal to the usable build area. The largest system, Maxum, offers a 23.6 × 19.7 × 23.6-in. (599 × 500 × 599-mm) build area. Titan offers a maximum part size of 16 × 14 × 16 in. (406 × 356 × 406 mm), and Prodigy Plus, the smallest, has an envelope of 8 × 8 × 12 in. (203 × 203 × 305 mm).

Using commercially available ABS adhesives, the bond strength on a fused deposition modeling part is suitable even in functional applications. In addition, these parts can be ultrasonically welded.

Powder-binder Printing

The ZPrinter 310 and Z406 offer a build envelope of 8 × 10 × 8 in. (203 × 254 × 203 mm), and the larger Z810 offers an envelope of 20 × 24 × 16 in. (508 × 610 × 406 mm). For each system, a small portion of the envelope is not available for part building. For example, the actual build envelope of the Z406 is approximately 7.5 × 9.5 × 7.5 in. (191 × 241 × 191 mm).

Like selective laser sintering, the porosity of the parts allows bonding agents to penetrate the part surface. As a result, when large parts are sectioned and later bonded, the bond strength is very good and may be stronger than the sections themselves.

COMPARISON OF OPERATIONAL PROPERTIES

There are aspects of the technologies' operations that affect the quality and usefulness of the rapid prototype. To fully define the similarities and differences between each system, the operational properties should be considered.

Time

It is unfair and unwise to compare the four systems on the sole basis of machine run time for the construction of a single prototype. Since the build time for each process is dependent on different parameters, time comparisons between technologies can be misleading. For example, some technologies are sensitive to the height of the part, while others are not; some require a pre-heat and cool-down cycle not required for other processes; and some are sensitive to the prototype's material volume.

To accurately compare the technologies for an evaluation, time studies should include representative samples of a company's requirements in a variety of configurations.

Additionally, the study should include all aspects of time, not just the run time of the system.

For a general presentation of build time, time studies of a test part are presented in *Figures 7-11* and *7-12*. *Figure 7-11* shows only the part construction time, while *Figure 7-12* shows the impact of additional time consumed before and after the part build.

Stereolithography

The time for the stereolithography process is dependent on many factors, including part volume, build height, build style, and material. In the time study of *Figures 7-11* and *7-12*, only one test part was built, therefore the time advantages of building multiple parts are not illustrated.

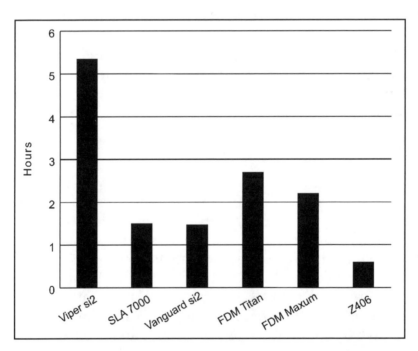

FIGURE 7-11. Build time data for the test part illustrated in *Figure 7-8*. Reported time does not include any preparation or post-processing time (Grimm 2002b).

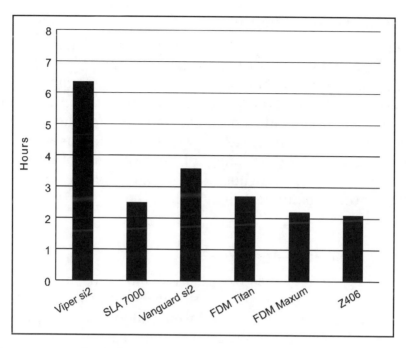

FIGURE 7-12. Elapsed time for the test part illustrated in *Figure 7-8*, which includes all preparation and post-processing time. Note: reported time does not include support removal (Grimm 2002b).

In the analysis, stereolithography produced the test part in 5.4 hours on the Viper si2 and in 1.5 hours on an SLA 7000. Both systems were run with layers of 0.006 in. (0.15 mm). For either system, the use of small, 0.001–0.002-in. (0.03–0.05-mm) layers would yield an increase of two to five hours. If the small spot option for the Viper si2 were used, the construction time could increase by an additional 150%.

An additional time consideration for stereolithography is the necessity of post-build operations. After constructing the prototype, the parts are cleaned in a solvent bath and cured in a UV oven. For the test part, this would add an additional hour to the process time. This time does not account for the support removal process.

Selective Laser Sintering

Selective laser sintering processing time is dependent on the same variables common to stereolithography, including part volume, build height, build style, and material. Like stereolithography, there are gains in speed when multiple parts are constructed in one machine run. For the test part in *Figures 7-11* and *7-12*, a Vanguard si2 produced the prototype in 1.5 hours using a layer thickness of 0.006 in. (0.15 mm). This does not include the pre-heat and cool-down cycles required. When added, the total process time is 3.6 hours.

Fused Deposition Modeling

Run times in the fused deposition modeling process are dependent on significantly different factors than the other technologies. In general, fused deposition modeling requires longer run times than the other technologies. For the Maxum, Titan, and Prodigy Plus, run times for the test part, constructed with 0.010-in. (0.25-mm) layers, range from 2.2–4.2 hours.

Run times for fused deposition modeling are dependent on the material volume of the part and support structures. Unlike the other technologies, the height of the prototype has no appreciable impact on time. Additionally, the material deposition rate, which is a function of the tip size, road width, and layer thickness, determines build times. The final factor that impacts build time is similar to that of CNC machining. Longer passes allow the extrusion head to reach maximum velocity, while shorter passes may not. Therefore, the geometry of a part on a given layer can play an appreciable role in determining build time.

Unlike stereolithography and selective laser sintering, fused deposition modeling run times do not vary by material. Also, the time per part is unaffected when constructing multiple parts on one machine run. Therefore, there is little gain in efficiency when building more than one part in a single run of the system.

The fused deposition modeling process does not require any additional time on the front-end or back-end of a build. Therefore, the time listed for the test part is the actual time from start to finish. This does not include support removal, which can approach 2–4 hours when using the dissolvable support structures of WaterWorks™.

Powder-binder Printing

Speed is one of the key advantages of powder-binder printing, as shown in *Figure 7-11*, which lists a build time of only 35 minutes for the Z406 with the plaster material. Build times for the powder-binder printing process are nearly constant and quite predictable. With little variance, time can be forecast with a consistent time per layer calculation. The key determinants of time include layer thickness and part height. While the *X-Y* footprint of the build does affect time—due to multiple print-head passes—this may not be a significant factor for many parts.

Material selection determines layer thickness and, therefore, build time. The starch materials build with 0.007-in. (0.18-mm) layers, and the plaster material uses 0.0035-in. (0.09- mm) layers. Since the layer thickness for starch materials is twice that of plaster, there are half the number of layers, which results in build times that are approximately 50% of those for plaster.

Before removal of the prototypes, the parts are typically allowed to remain in the machine for 1–3 hours. For the test part, this brings the total processing time to 1.6 hours. This allows the binder to dry and harden. The length of drying time is a function of the material—starch takes longer to dry than plaster—and the wall thickness of the part. In some cases, the part will be allowed to further dry after removal from the system.

Support Structures

Support structures may have a measurable impact on time and prototype quality. With the specification of support structures, machine time and labor time will increase. Additionally, when removing the supports, surface finish and feature integrity may be diminished.

Stereolithography

Stereolithography requires support structures to success-fully build a prototype. Support structures are thin ribs placed at approximately 0.25-in. (6.4-mm) intervals in a checkerboard pattern. Supports serve two purposes. First, they rigidly attach the prototype to the build platform. This anchors the part to prevent it from shifting while the plat-form dips into the liquid bath and the blade sweeps the surface of the resin. Second, the supports attach to any downward facing surface to fix the feature in place. Without this support, a single layer of cured material would sway or deform within the liquid environment.

To complete the prototyping process, the supports must be removed. This post-processing can affect the accuracy of the geometry. When manually removing the support, toler-ance deviations will result from the removal of too little support or too much part. Additionally, the properties of some materials make the stereolithography models suscep-tible to chipping and breakage of small details. Accessi-bility to support structures is also a concern. Should a part have internal cavities with little or no access, the supports will remain in the prototype.

Selective Laser Sintering

Selective laser sintering does not require any form of support structure. The powder that surrounds the sintered material acts as a fixture by encasing the prototype in a "cake." When a build is complete, brushing, vibrating, or air blasting the cake will expose the prototype to complete all required processing.

The only quality consideration for selective laser sintering is the ability to fully evacuate the residual powder from areas that may not be readily accessible.

Fused Deposition Modeling

As with stereolithography, fused deposition modeling requires a supporting structure to form a base on which to mount the part and to support any overhanging features. At the interface with the platform, a solid layer of supporting material is laid down. Above this solid layer, supports with 0.020-in. (0.5-mm) and 0.150-in. (3.8-mm) gaps are deposited.

Fused deposition modeling offers two styles of supports: break-away support structures (BASS™) and water-soluble support structures (WaterWorks™). BASS supports are manually removed by stripping them from the part surface. While they do not mar the part's surface, consideration must be given to the accessibility and proximity to small features. WaterWorks uses a soluble support material that is dissolved in a water/solvent solution. Unlike BASS, the supports can be located in deeply recessed regions of the part, or in contact with small features, since mechanical removal is eliminated.

Powder-binder Printing

Like the selective laser sintering process, powder-binder printing does not require a support structure. As the part is constructed, it is surrounded by unused powder that supports and retains the part during the build process. After part construction, the excess powder is brushed away from the part surface. Due to the properties of the prototype, care should be taken to remove the excess powder without damaging small features.

Post-build Processing

Following the construction of a rapid prototype, each technology requires additional processing. In general, these actions include cleaning, curing, and support removal.

Stereolithography

Following the build of a stereolithography prototype, it requires cleaning, curing, support removal, and light benching. When extracted from the system, the prototype is coated with uncured resin. In a cleaning operation, the excess material is stripped from the part surface. While some users perform this step manually, many use solvent tanks. Once cleaned, the prototype is placed in a UV oven to complete the curing. After curing, the support structures are removed from the part.

Since support structure removal leaves visible surface defects, the supported surfaces of most stereolithography prototypes are lightly sanded. At this point, the prototype is ready for distribution. However, users often bead blast the prototype to improve cosmetic appeal. For applications such as patterns or painted models, the stereolithography prototypes require additional finishing. To remove stair stepping and to smooth vertical surfaces, the parts are further benched.

Selective Laser Sintering

When removed from the system, the selective laser sintering prototype is encased in unsintered powder. This excess material is removed to yield the finished prototype. In most cases, this process is completed quickly. However, for prototypes with difficult to reach areas, material removal may be time consuming.

Unlike stereolithography, no additional processes are required when producing plastic prototypes. However, users often infiltrate the surface of the prototype with compounds such as cyanoacrylate. This operation seals the prototype and provides a sandable surface. Without the infiltrant, sanding is unable to yield a smooth surface since the underlying material has the same porous characteristics.

For metal prototypes and tools, a furnace cycle is required. To burn off the binder and infiltrate with bronze, the parts are heated in a furnace over a 24-hour period.

Fused Deposition Modeling

As with stereolithography, after the build is complete, the support structure of a fused deposition modeling prototype must be removed. As discussed previously, fused deposition modeling offers two types of supports: BASS and WaterWorks. With BASS, the supports are manually stripped from the part surface. WaterWorks, on the other hand, uses a soluble material that is dissolved.

For both supporting methods, no visible surface mars remain. Therefore, there is no requirement for additional sanding. However, to overcome the limitations of surface finish, many users elect to perform secondary operations. With the toughness of the ABS and polycarbonate materials, sanding may be somewhat laborious. Users often report the use solvents or adhesives as a one-step process or in preparation for sanding. Commercially available, these agents include Weld-On®, ABS glue, acetone, and two-part epoxies.

Powder-binder Printing

For short-lived concept models, many users elect to use powder-binding printing prototypes as-is from the machine. However, when the prototype is used for form and fit or other advanced applications, infiltration is necessary. When infiltrating and sealing with wax, an automated waxer can be used. For all other infiltrants, the materials are applied manually. Depending on the selected material, the prototype may require a drying period and, in some cases, an oven cycle to fully cure the infiltrant.

COMPARISON OF APPLICATIONS

Combining the physical and operational properties of each technology, appropriate applications can be chosen. Although each technology can be used for all applications in the design and manufacturing process, the choice of any one may not be the strongest solution in each instance. This comparison of technologies by applications does not indicate that a given system is best or worst for the application. That is a decision that must be made in light of the requirements of a given project. Instead, this comparison offers a relative positioning of the technologies with the typical requirements of a general application.

Concept Models

Short-lived concept models are used for early design validation or communication. This application of rapid prototyping could also be called a visual aid. Much like a printed first draft of a manuscript, the concept model is used for proofing and is quickly discarded. Multiple design iterations are often reflected through concept models. An example of a concept model is shown in *Figure 7-13*.

FIGURE 7-13. Example of a concept model. *(Courtesy Z Corporation)*

Stereolithography

Concept modeling is generally perceived to be fast and inexpensive. While stereolithography offers reasonable speed, the expense of the technology and the operational demands do not position it well as a concept modeler. However, stereolithography offers many benefits appropriate to this application. Surface finish, accuracy, and replication of detail make this technology well suited for models used for display and communication purposes.

Selective Laser Sintering

With its key advantage being functional material properties, selective laser sintering is not commonly applied to concept modeling. Like stereolithography, the expense and operational demands position the technology for higher-level applications. Yet, even in a show-and-tell environment, material properties can be a concern. When illustrating a product design, it can be unnerving when a snap fit fails or a prototype breaks. For these reasons, the durable material properties available in selective laser sintering can be highly beneficial in a concept-modeling environment.

Fused Deposition Modeling

Many fused deposition modeling users treat the technology as a design peripheral, especially when using the Dimension (3D printer) product line. As such, the technology becomes another tool linked to and driven by the CAD system for the purpose of interrogating and validating design concepts early in the process. While fused deposition modeling may not offer the speed expected from a concept modeler, it offers a combination of benefits important for concept modeling and visualization applications. These strengths include ease of operation, accuracy, material properties, color, and the elimination of manual part finishing.

Powder-binder Printing

By far, the most common application of powder-binder printing is concept modeling. Considering the low cost of the technology and materials, the small footprint, office environment operation, ease of operation and, of course, the speed, these devices are ideally suited for concept modeling. Additionally, the limitations of the technology are not significant in the early evaluation of a new product design. When using the prototype as a visual aid, accuracy, surface finish, and material properties may not be critical to the application.

An additional consideration is the application of color. To mimic the overall appearance of a product, powder-binder printing can produce a realistic representation of the final product. Furthermore, color can be applied to a prototype to illustrate, in three dimensions, the results of an FEA or mold flow analysis.

Form, Fit, and Function Models

Design analysis is perhaps the most important role of a rapid prototype. The ability to quickly verify a design prevents the investment of time and money in a poorly conceived project or component. After the concept has been defined, the design progresses into its mechanical definition. At this stage, the prototype is used to confirm that parts properly mate and allow ample space for components. The other aspects are confirmation of the overall design form and its ability to function as specified. *Figure 7-14* illustrates an example of such a model.

Stereolithography

Stereolithography excels in producing form and fit models, but the technology is limited in functional applications. For form and fit models, the combination of detail, accuracy, and surface finish makes stereolithography a widely used tool. However, the limitations in material properties often override these advantages when the technology is considered for functional application. Yet, recent advances in material development have opened the door to light to moderate functional testing with stereolithography prototypes.

Selective Laser Sintering

Like stereolithography, there are trade-offs between strengths and limitations in the application of selective laser sintering prototypes. However, the roles are reversed.

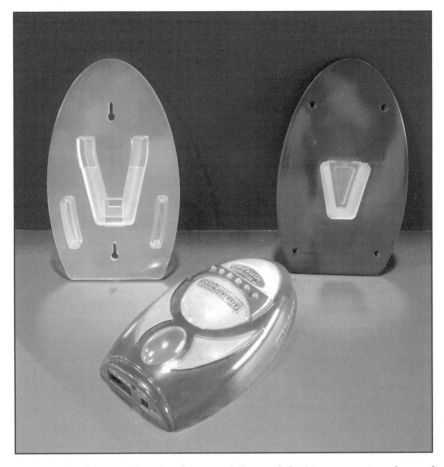

FIGURE 7-14. Example of a form and fit model. *(Courtesy Accelerated Technologies)*

Selective laser sintering's greatest strength is the functionality of its materials. Strong, robust, and insensitive to many environmental factors, these prototypes are well suited for functional analysis of plastic and metal parts.

While the technology is used for form and fit analysis, the surface finish, accuracy, and detail replication detract from its usefulness in this application.

Fused Deposition Modeling

Fused deposition modeling prototypes are frequently used for form, fit, and function analysis because there are not significant trade-offs with the other technologies. The technology offers strong and tough prototypes with the detail, accuracy, and machinability required for functional analysis of molded plastic parts. Although an unfinished part may not have the surface finish of a production part, many do not find this to be an obstacle for this application. Additionally, surface finish is often secondary to other factors such as dimensional stability, heat resistance, and chemical inertness.

Powder-binder Printing

While the selection of the proper infiltrant may allow the application of powder-binder printing to form, fit, and function analysis, often the demands of the application exceed the deliverables of the process. The advantage of speed is overshadowed by the deliverable accuracy and surface finish. Although post-processing can improve on these aspects of the prototypes, the time and effort associated with the process and the infiltration of the parts further detract from the primary advantage of the technology, speed.

Often, users will use powder-binder printing for all of their concept modeling needs. After finalizing the design concept, they then turn to another technology for the form, fit, and function prototype.

Patterns

When rapid prototyping is not suited for the desired quantity of prototypes or it does not offer the desired material properties, the prototypes are frequently used as a pattern for the formation of a mold. One example is the use of rapid prototyping to create a pattern for rubber molding. The pattern has liquid rubber compound poured against it. After the rubber sets, the pattern is extracted, yielding a cavity into which liquid, thermoset plastic is cast.

Other pattern-based applications are found in metal-casting applications. *Figure 7-15* shows a pattern, and the resulting casting, from the investment casting process.

Stereolithography

Tooling patterns are tolerant of a wide variety of material properties. This tolerance promotes the advantages of stereolithography. A tool and the parts it produces can only be as good as the underlying pattern. Therefore, the surface finish, accuracy, and detail replication available from stereolithography make it a preferred method for pattern generation.

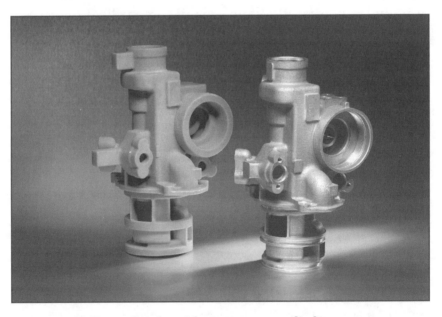

FIGURE 7-15. Example of rapid prototypes applied to pattern generation. The pattern (left) was used in the investment casting process to form the metal valve body (right). *(Courtesy Accelerated Technologies)*

With the QuickCast® build style, stereolithography patterns are also used in the investment casting process. QuickCast produces a hollow part with an internal lattice that supports bounding surfaces. While successful in investment casting applications, the properties of the stereolithography resins require modification to the traditional foundry processes.

Selective Laser Sintering

Pattern generation is one of the weaker applications for selective laser sintering. The strengths of the technology are not overly applicable. Meanwhile, the limitations of surface finish requires a significant amount of part benching to produce a suitable pattern.

However, selective laser sintering is often applied to investment casting patterns. Using a polystyrene material, the technology delivers a pattern well suited for the burn-out process in investment casting. Unlike the stereolithography QuickCast solution, these patterns do not demand modification to a foundry's standard process.

Fused Deposition Modeling

As with the other rapid prototyping technologies, fused deposition modeling can be successfully used as a pattern generator. However, consideration must be given to the surface finish and time required to bench a part to master pattern specifications.

Wax and ABS patterns constructed from the fused deposition modeling process have proven to be suitable for investment casting. Like selective laser sintering, the patterns are processed with minimal modification to the standard foundry processes.

Powder-binder Printing

When using rapid prototyping for pattern generation, speed is often secondary to overall prototype quality. This limits the application of powder-binder printing for pattern building. While there are examples of powder-binder printing being used for pattern generation, in many cases it is not a primary application for the technology.

Rapid Tooling

Bridging the product development and manufacturing processes, rapid tooling is used to create prototypes, short runs of production items, and even full manufacturing runs. Used indirectly as a pattern generator or directly to create a tooling insert, the rapid prototyping device crafts the mold for processes like injection molding, investment casting, or reaction injection molding (RIM).

Stereolithography

The stereolithography process is commonly used for indirect rapid tooling applications. Building on the process' strengths as a pattern generator, tooling is cast or formed from the stereolithography pattern.

While there have been successful uses of stereolithography for direct applications, the material properties of the tooling inserts are suited only for injection molding of extremely low quantities of prototypes in easily molded thermoplastic materials.

Selective Laser Sintering

Selective laser sintering has a distinct advantage in the direct production of tools. Designed for tooling applications, LaserForm, the binder-coated, stainless-steel powder, delivers a metal core and cavities with a robust composition of 60% stainless steel and 40% bronze. Tooling constructed by selective laser sintering is often used for injection molding applications. Although many users have achieved significant time and cost savings with LaserForm tools, there are limitations in surface finish and accuracy. Often, secondary machining is performed to improve these characteristics.

Fused Deposition Modeling

Using thermoplastic materials, the fused deposition modeling technology is not suited for direct rapid tooling application. While the materials are strong, they do not offer the thermal properties required for tooling. Additionally, as indicated in the pattern application, the surface finish of the fused deposition modeling parts limits the application of the technology as an indirect solution.

Powder-binder Printing

For direct rapid tooling applications, powder-binder printing is used for the creation of molds for the foundry industry. Metal casting applications like investment casting, sand casting, or plaster mold casting use molds with properties similar to those of powder-binder printing. In these applications, the mold is sacrificial; it is destroyed once the cast metal has hardened. As a result, these casting applications use molds with materials that can retain the molten metal and withstand high temperatures, yet be easily broken away from the metal part once it has solidified.

It is important to note that there are other MIT licensees offering systems based on the 3DP process to address the needs of the rapid tooling industry. For example, one technology uses metal powder joined by the binder. After construction, the tool is fired in a furnace to burn off the binder and infiltrate the tool with metal.

Rapid Manufacturing

Rapid manufacturing is a concept of producing finished goods on demand without the need for tooling, molding, or machining. Applications already exist and more are likely to develop. Rapid prototyping has fueled an interest in short-run manufacturing where economic order quantities could be as small as a single unit. This application requires that the part meet functional specifications in many areas.

Stereolithography

As with rapid tooling, when indirect approaches are considered, stereolithography can be a rapid manufacturing device. For these manufacturing processes, stereolithography creates a pattern used to create tooling for the production of end-use items. In these applications, the combination of accuracy, surface finish, and detail position the technology as a strong solution.

In some cases, stereolithography has been used as a direct rapid manufacturing solution. While the material properties available in stereolithography are not suited for many end-use items, some parts can accommodate the material limitations while benefiting from the other quality characteristics.

Selective Laser Sintering

Building on the functionality of materials, selective laser sintering can be successfully applied to the direct manufacture of production parts. Surprisingly, one of the most stringent and demanding applications, aerospace, was one of the first rapid manufacturing applications. For several years, selective laser sintering has produced parts used in the space shuttle and the international space station.

The limiting factors for the application of selective laser sintering in rapid manufacturing applications include surface finish and dimensional accuracy. For tight tolerance parts or visible, cosmetic components, this process may not be suitable. Yet, there are many mechanical components that do not demand this level of quality. It is for these that the technology is best applied.

Fused Deposition Modeling

With the accuracy and material properties available from fused deposition modeling, it is positioned to address rapid manufacturing applications. While an unfinished fused deposition modeling part would have limited use in visible, cosmetic applications, this is not an obstacle for internal components or those that do not require aesthetic appeal. For rapid manufacturing applications, the part construction time for fused deposition modeling could also be an important consideration. Yet, as some users have proven, the run times for a few parts are significantly less than the total time required to produce tooling and parts.

Powder-binder Printing

While powder-binder printing has the speed desired for rapid manufacturing applications, it lacks the quality and material properties required for the direct production of metal and plastic parts. However, rapid manufacturing can apply to much more than mechanical components for consumer and industrial goods. Several MIT licensees have applied powder-binder printing to the direct manufacture of finished goods. In one case, complex, custom ceramic filters are printed directly from the rapid prototyping device. In another case, a company replaces the binder with custom blends of pharmaceutical drugs. Using the starch material as a carrier, this company prints custom formulated tablets for drug therapy.

CONCLUSION

The detailed description of the four technologies is a starting point for a technology comparison. There are many other details of the technologies that should be considered when selecting a technology. This is especially true when considering a system purchase. For in-house operation, the technology comparison should extend to the requirements for implementation, daily operation, and cost of ownership.

There are dozens of technologies from which to choose. While only four technologies were compared and contrasted in this chapter, the information offered will serve as a valuable baseline for information gathering. Presentation of the comparative information on the four technologies may serve as a guide to the development of a technology evaluation.

REFERENCES

Grimm, Todd. 2002a. "Fused Deposition Modeling: A Technology Evaluation." Edgewood, KY: T. A. Grimm & Associates, Inc.

——. 2002b. "Stereolithography, Selective Laser Sintering, and PolyJet™: Evaluating and Applying the Right Technology." Erlanger, KY: Accelerated Technologies.

CHAPTER 8

Developing a Justification

With the applications, advantages, and benefits of rapid prototyping, it is obvious that the technology can be a powerful tool for most companies in most industries. With some machines selling for under $30,000, acquiring a system can be an attractive proposition.

Yet, the justification for the purchase and implementation of rapid prototyping is not always straightforward, clear cut, and easy. While systems sell for as little as $30,000, there are other expenses and burdens to be included in the cost analysis. Although the benefits of rapid prototyping may be obvious, inherent, and intuitive, quantifying the advantages can be challenging. Many of the benefits of rapid prototyping are intangible and hard to measure.

A successful justification that leads to powerful long-term results demands the identification of all costs and burdens, and determination of benefits in terms that can be measured and valued. An evaluation and justification must also consider the two methods of securing prototypes: in-house operation or subcontracted services.

IN-HOUSE OPERATION VERSUS SERVICE BUREAUS

Before or during a rapid prototyping review, it is important to evaluate system ownership versus the use of subcontracted services. Each has advantages and disadvantages, but both offer the benefits of access to rapid prototyping.

Since the earliest days of rapid prototyping, businesses have offered the development of prototypes as a service. Initially these companies, called service bureaus, allowed organizations access to an expensive and unproven technology. The service bureaus offered the benefits of rapid prototype development without the high upfront expense and significant risks of implementing a new technology that was relatively unproven and poised for rapid obsolescence. In the early years, the service bureaus also offered file conversion to bridge the operational gap between 2D data and 3D computer-aided design (CAD) models, making rapid prototyping available to the large number of companies that had yet to make the transition to solid modeling.

Hundreds of service bureaus operate around the globe. As the industry has changed and technology has progressed, these companies have modified the role they play and the services they offer to deliver advantages beyond the reduction in initial expense and risk of ownership. In doing so, service bureaus continue to be a viable option for many companies even with the availability of low-cost rapid prototyping systems. In fact, some seasoned rapid prototyping veterans advise prospective rapid prototyping users to first use the services of such companies before launching an evaluation and implementation of rapid prototyping technology.

Rapid prototyping can be difficult to fully evaluate, since there is a limited body of publicly available information. This is most evident when trying to determine the cost of operation and any limitations of the technology. With dozens of available systems and limited information, it can be challenging to select the most appropriate device for the current needs. For this reason, many companies elect to use service bureaus prior to a system purchase. The use of the service bureau allows the evaluation of multiple technologies and materials with minimal expense and risk. It also establishes a baseline for financial projections and justifications.

Prior to, or concurrent with, a system evaluation and implementation (as detailed in Chapter 9), weigh the pros and cons of ownership versus use of a service bureau. As with any process that requires capital investment, operational overhead, and staffing, it is important to consider acquisition and operation of a rapid prototyping system versus outsourcing the work to a qualified vendor. The justification for either approach will be made with consideration of system utilization, expected benefits, and total expense.

ECONOMIC JUSTIFICATION

Justifying the implementation of rapid prototyping can be difficult even with the availability of low-cost systems. The challenge arises for two reasons: the total cost of acquisition and operation and the difficulty of proving a return on the investment.

With system prices of $30,000–800,000 and total start-up cost of $50,000 to over $1 million, many organizations assume rapid prototyping is out of their reach. When adding annual operating expenses—materials, maintenance, power, labor, training, and support—the financial justification becomes even more difficult. This is compounded by the difficulty of justifying the technology based on factors that are hard to measure: cost avoidance, quality improvement, and time reduction. If these items cannot be measured and quantified, a financial justification is less likely to succeed.

Many of the benefits of rapid prototyping arise from what is avoided rather than what is eliminated or improved. Unlike the purchase of a new machine tool, which is justified on faster throughput and decreased cost per piece, rapid prototyping often offers the benefit of avoiding mistakes, improving quality, or decreasing time to market. Since mistakes, quality issues, and product development cycle times will be one-time occurrences for a specific product, there are no direct comparisons available, only historically similar situations. Even the decrease in time to build a prototype may not be satisfactory for justification unless the value of that time translates into a beneficial business outcome.

For those who have experience with the justification of a transition from 2D CAD to 3D solid modeling, the similarity to rapid prototyping is clear. Before the purchase, it is hard to translate better and faster design work into real, measurable numbers. For both 3D CAD and rapid prototyping, the benefits are more apparent after system installation because there are real examples of the time and money saved on specific projects.

Strategies

One strategy for justification is to address the financial performance of the company and illustrate how rapid prototyping will produce an impact on profitability. By addressing the high-level financial concerns rather than the value on an operational level, the justification is often easier to prove. This strategy also negates the difficulty in quantifying a measurable result.

Since an expenditure will require management approval, quantifying the financial gain in terms of sales revenue or profit, management's key performance measures, is often well received. A hypothetical example of justification strategy is on the next page.

In this example, rapid prototyping could be justified with only a 1.9% reduction in time to market. With the impact on design time and design revision, it is often easy to accept that rapid prototyping can yield this small time decrease.

Total acquisition and first-year expense for rapid prototyping: $500,000

Projected new product sales (first year): $25,000,000
Projected gross profit: $12,000,000
Daily gross profit: $48,000

Where:

Break-even point = Expense ÷ Daily gross profit

So:

Break-even point = 500,000 ÷ $48,000/day
Break-even point = 10.4 days

If:

Time to market (historical) = 18 months

Then:

*Break-even point = (10.4 days ÷ 30 days/month) ÷ 18 months ×
100%*

So:

Break-even point = 1.9%

For those companies that have used service bureaus, another strategy is to base a justification on the savings and additional benefits of bringing the technology in-house. Using service bureaus prior to a system purchase yields an enormous amount of data that supports the justification process. Most notable are the cost of the prototypes and the average delivery time. Using the actual expense and delivery time, a justification can compare the projected cost and delivery time of an in-house operation versus subcontracted services. Additionally, the use of the service bureau offers insight into the operational demands, cost structure, strengths, and limitations of a given technology.

Of course, there are many other methods for financial justification of rapid prototyping. Alternatives include efficiency, throughput, fully burdened hourly cost, and enhanced capability. One factor that should not be ignored is the reduction in demand for labor and skilled personnel. Companies are leaner today, and the pool of skilled trades-people continues to shrink. Rapid prototyping could be justified as a tool that supports work force reduction initiatives or as a solution to the difficulty in hiring skilled CNC machinists, toolmakers, or CAM programmers.

A key to a successful justification is the determination of the critical needs of the organization. If labor cost is the issue, justify rapid prototyping by showing decreased labor demands. If time to market is the sweeping initiative, illustrate how the combination of computer-aided design/computer-automated manufacturing (CAD/CAM) and rapid prototyping can slash weeks from the design cycle and product launch. If cost control is the biggest challenge, prepare a justification that clearly demonstrates how the system can reduce expenses and by how much. As always, combining multiple benefits into the justification will demonstrate higher return. While there are many other factors beyond cost that may justify a rapid prototyping implementation, the starting point should always be what management is concerned with the most, the bottom line.

All of the aforementioned strategies make one major assumption. Each assumes that all costs, burdens, and limitations can be accurately defined. Without investigation and diligence, this assumption may be flawed, since many costs associated with rapid prototyping ownership are difficult to obtain. For a strong justification and long-term success, these hidden costs should be uncovered.

HIDDEN COSTS (Wohlers and Grimm 2002)

Even with an extensive evaluation or the use of service bureaus, there may be areas of time and cost that are not recognized until after the purchase and implementation of a rapid prototyping system. These often are unidentified costs of operations, labor, capabilities, and expenses.

One example that many stereolithography users have experienced relates to part finish and quality. With the competitive pressures that service bureaus face, each works diligently to maintain some advantage over their competitors. One such area is in part finishing (often referred to as benching). To satisfy the demands of their customers, service bureaus that use stereolithography most often deliver a prototype that is smooth and free of surface defects. Some offer a standard finish that is ready for a coat of paint. When a company shifts from contracting with service bureaus to in-house operations, it is often surprised to find how much time and labor is required for high finish levels. In effect, the service bureau buffers the customer from system limitations and the real cost of operation. This example represents just one of the many aspects of rapid prototyping operations where data and information are not readily and publicly available. Many companies do not disclose their experiences and financial data to protect proprietary information. The result is that there is a substantial information gap regarding the true cost of rapid prototyping system implementation and operation.

As will be discussed in Chapter 9, it is critical to conduct a thorough review of the capabilities and limitations of rapid prototyping systems when making a selection. A cursory evaluation may lead to the acquisition of a technology that poorly supports the needs of the organization or one that demands more resources than the organization can bear. One of these resources, the financial one, is critical in both the justification of the technology and its long-term success.

With a multitude of technologies and individual systems, it is inappropriate to attempt to detail all of the financial aspects of each system. However, in reviewing key aspects of four leading technologies, a strong foundation for information gathering and budgeting can be formed. Here, some primary, but often overlooked, areas of rapid prototyping expense are described.

System Expense

Rapid prototyping machines have a wide range of options, capabilities, and prices. With advertised prices of $30,000–800,000, the purchase of the equipment is a quantifiable expense. Yet, the purchase of an inappropriate device can add undue financial burden to the justification and long-term operation. It is unwise to assume that the least expensive technology will do the job. Likewise, it is equally unwise to conclude that the organization must have the very best (most expensive) to satisfy its needs. To accurately determine which price range will satisfy the organizational demands, review Chapter 9 to gain an understanding of how to identify the technologies (and their associated prices) that will deliver the desired results.

A long term, hidden cost with regard to the system purchase is the risk of obsolescence. For technologies that are early in their life cycle, such as rapid prototyping, rapid advancement and improvements present the threat of owning a technology that is no longer viable. However, the risk of obsolescence should not paralyze a rapid prototyping evaluation. Historically, most systems have not suffered outright obsolescence. For example, early SLA® 250 machines, circa 1990, are still in operation throughout the world.

To avoid obsolescence and maintain system performance at currently accepted levels, account for an upgrade path. Hardware and software upgrades are common for rapid prototyping devices. And while these enhancements improve performance and output quality, they can be relatively expensive. Historically, hardware upgrades have been in the $5,000–50,000 range.

Facility Improvements

Although vendors offer detailed information on facility requirements, the specifications may arrive after the purchase if they are not requested during the evaluation. By simply asking for this data, the hidden cost becomes a quantifiable expense. Ignoring this area can lead to shock and disappointment. While some systems may require only a little space and a wall plug, others may demand extensive facility modifications that could exceed the purchase price of the system.

Since most rapid prototyping systems are best suited for a lab environment, not the shop floor or office, facility modification can be extensive. For these operations, the obvious expense is in construction of the lab. Yet, this is not the only expense. Depending on the system selected, other alterations can include electrical, plumbing, and HVAC work. High-voltage power supplies, waterlines, gas lines, temperature controls, and humidity controls are some of the more common site modification requirements. *Figure 8-1* illustrates such an environment.

For some systems, it is best to consider an isolated environment. The processes can produce airborne dust and vapors that at best are unpleasant and at worst an unacceptable hazard by corporate standards and OSHA codes. When isolating the lab, there will be additional expense for ventilation, air handling, and possibly dust collection.

An often-overlooked expense is the internal facility charge. If the organization allots an overhead expense to departments based on the percentage of total square footage used, secure this number and add it to the justification and operating budget.

Finally, the facility modification may not be limited to the rapid prototyping lab. Prototype cleaning and finishing will require a shop floor or model shop environment. If a suitable area does not exist, the expense for this facility modification, including electrical, plumbing, and HVAC, should be budgeted.

FIGURE 8-1. Many rapid prototyping systems are best suited for a lab-like environment that is isolated from office and shop-floor areas. *(Courtesy Accelerated Technologies)*

Staffing

As noted earlier, rapid prototyping offers the benefits of demanding little labor and operating in an unattended mode. However, this does not mean labor expense can be ignored in the financial calculations. On the contrary, staffing can be an appreciable component of the budget.

For basic applications where the rapid prototyping system does not require a dedicated operator, it could be assumed that individual designers and engineers will process, build, and clean their own prototypes. However, while this scenario does not demand a direct labor expense, it will impact the efficiency and productivity of the individuals. For some organizations, it is important to include a calculation of this indirect labor expense.

For most systems, it is more likely that a dedicated rapid prototyping staff will be required. Depending on the system and application, the staff could be a single employee or a complete department. Labor will be required for data preparation, part construction, part cleaning, and benching. Additionally, responsibilities will extend to departmental management, maintenance, and customer service.

Since most rapid prototyping systems have yet to achieve a push-button mode of operation, staff will require classroom training and hands-on experience. The training time and expense should be considered in the budget. Additionally, consider the overtime expense for non-exempt (hourly) employees. Efficient rapid prototyping departments do not operate between the hours of 8:00 a.m. and 5:00 p.m. With the demand for rapid delivery of parts, working during evenings, weekends, and even holidays is likely for this department.

In terms of staffing, the first RP machine will be the most expensive. As an operation grows, it is common for one skilled technician to operate multiple machines, as shown in *Figure 8-2*. As a result, the multiple machine environments will have a lower percentage of labor to total operational or prototype cost.

Material Costs

Rapid prototyping materials are expensive, and the cost swells with waste, scrap, inventory, and material conversions.

Photopolymers, such as those for stereolithography, sell for approximately \$90/lb (\$41/kg). Although the price can vary by product order quantity, these materials are far from being a commodity. At this price, the initial cost of filling a small stereolithography system, like the Viper si2™, is nearly \$4,000. Larger systems, like the SLA® 5000 or SLA® 7000, cost nearly \$40,000 to fill. Since stereolithography

FIGURE 8-2. In the distance, an operator can be seen preparing a selective laser sintering system for a build. As the number of systems increases, labor costs can decrease since one technician can operate multiple machines. *(Courtesy Accelerated Technologies)*

systems must be replenished prior to each build, an inventory expense must be considered. For example, each stereolithography material will require an on-hand inventory of $3,000 worth of material or more.

In most cases, unused material cannot be completely reclaimed. For stereolithography, the cumulative exposure to ultraviolet (UV) energy can cause an entire vat of resin to become unusable. With selective laser sintering, waste is generated with each build, because the process requires a ratio of 33–50% virgin material to recycled powder. Since the ratio of unsintered powder to part volume is always high (typically 10:1), much of the used powder will never be reclaimed. This creates a stockpile of expensive, unusable material for disposal.

Bad builds and damaged parts also add to scrap material rates. With the many variables associated with producing a good prototype, it is unreasonable to expect 100% performance. Although the scrap rate will vary, it would be wise to plan for a 10% loss. With all of these factors, expect an allocation of 20–30% of budgeted costs for materials.

Supporting Equipment

Rapid prototyping can require additional equipment for both the front-end and back-end processes.

The file processing, transfer, and archival needs of a rapid prototyping operation may require software, hardware, and networking not supplied by the rapid prototyping vendor. These front-end systems should be established to handle high volumes of files, some of which can be quite large. Although the rapid prototyping systems include the necessary software to analyze and process STL files (see *Figure 8-3*), many elect to purchase third-party programs that offer advanced tools or greater processing flexibility.

For back-end systems, allot an appropriate amount to equipment for cleaning and benching operations. For some technologies, the expense of the supporting equipment can approach or even exceed that of the rapid prototyping system. When considering all technologies, the list of back-end equipment requirements and options is too extensive to discuss in detail. However, key areas of consideration include part cleaning equipment (UV ovens, wash tanks, air compressors, downdraft tables, and workstations) and benching equipment (hand tools, power tools, workstations, spray booths, and machine tools). Examples of back-end equipment are shown in *Figures 8-4* and *8-5*.

Maintenance

Repairs and maintenance, as some users find, can be quite expensive for rapid prototyping equipment. Typically,

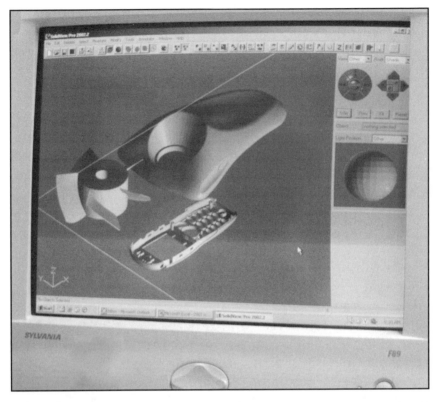

FIGURE 8-3. Front-end system requirements include computer hardware and software programs for processing STL files.

vendors will offer full-service maintenance contracts. Some offer a reduced rate for contracts that take into account reasonable time and material repairs. Additionally, software maintenance, support, and upgrades can be included in a contract or purchased separately.

A full-service maintenance agreement for a stereolithography system can cost nearly $75,000 a year. For this price, all software, service, repairs, and replacement lasers are covered. A one-year service agreement for a fused deposition modeling system is approximately $19,000. As a general rule, a budget amount of approximately 10% of the

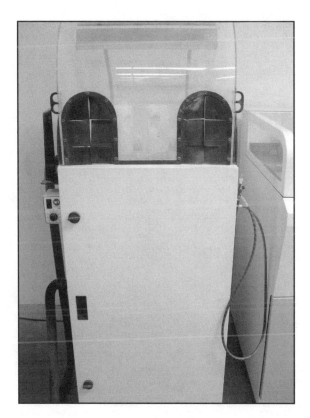

FIGURE 8-4. Once a prototype is constructed, it is then cleaned and finished. The de-powdering unit for a powder-binder printing system uses compressed air to blow excess powder from the prototype surface. *(Courtesy Fisher Design)*

purchase price of the machine should cover one year of full-service maintenance.

For those who choose the time and materials support option, maintenance cost can be reduced if repairs and replacements are required at a reasonable level. When selecting this option, plan on expenses for routine calibration, preventive maintenance, software upgrades, component failures, and consumables.

Many of the components in an RP machine will wear out or fail. A key consideration is the replacement frequency of

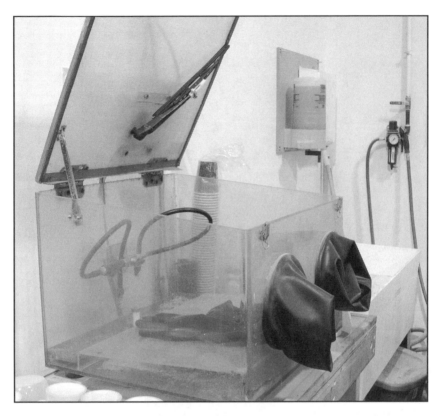

FIGURE 8-5. Back-end operations require equipment that varies with each technology. For some, a waterjet cabinet is used to spray residual material and support structures from the part surface. *(Courtesy Accelerated Technologies)*

a laser, print head, or nozzle. None of these items lasts indefinitely. For example, a solid-state laser for stereolithography has a warranty for 5,000 hours of operation—seven months of use running at 70% of capacity. At this operational rate, there would be a minimum of one new laser a year at a cost of $25,000–40,000.

Software maintenance can often be quite expensive too. When using the selective laser sintering process, software maintenance costs $6,000 per year per machine.

In general, service calls, maintenance contracts, and replacement parts are expensive. With relatively few systems in the field, there is a small base upon which to amortize parts, service, and support. As a result, maintenance is often expensive when compared to other products. There are few options for cost reduction. Many system components are manufactured specifically for the technology, so these must be purchased from the vendor.

Cleaning and Finishing

All rapid prototyping technologies will require some cleaning prior to part delivery or benching. The most common aspects of part cleaning are the removal of residual material and support structures.

Processes such as stereolithography and fused deposition modeling require the removal of support structures. This can be tedious and time consuming, and it may require special tools and equipment. Labor expense can be reduced with options such as WaterWorks™, a dissolvable support structure for the fused deposition modeling process. Yet, the reduction in labor cost may be offset by additional expense for equipment and chemicals.

Powder-based processes, such as selective laser sintering and powder-binder printing, omit the need for support structures. However, these processes require the removal of residual powder from the surface of the part. To do so requires labor and equipment expense. For selective laser sintering, loose powder is removed with compressed air and hand brushing while working over a downdraft table.

In the case of stereolithography, it is important to remove all uncured resin prior to the UV post-cure. Typically, an agitating tank filled with solvent, such as tripropylene glycol methyl ether or propylene carbonate, is used for this process.

When the prototypes require a high level of finish (for example, for tooling patterns or photo-quality models), staffing requirements include skilled and experienced technicians to bench and paint the prototypes. These processes also will require ample supply of consumables— sandpaper, gloves, adhesives, paint, and primer. And, as described earlier, the supporting equipment may represent a sizable expense.

Miscellaneous

Some of the materials must be treated as hazardous chemicals, so protective equipment may be required. Items may include gloves, lab coats, eye protection, and respirators. For hazardous materials, the cost of disposal of contaminated water, unused resin, and cleaning solvents will also be a budget consideration.

ADDITIONAL CONSIDERATIONS

While there is a significant cost for equipment ownership, there are many powerful advantages. As some companies have found, payback periods can be measured in months, not years. Even if organizational demands require one of the more expensive solutions, rapid prototyping can be, and often is, a good investment. Many organizations worldwide have had success with their RP installations and have even purchased a second, third, and even tenth machine. But before investing in rapid prototyping, consider the real costs, the true justification, and the advantages of in-house versus subcontracted services.

There are many other nonfinancial factors in the decision to use rapid prototyping and whether to build an internal department or rely on service bureaus. In some cases, the final decision may be as simple as adhering to corporate philosophy; some believe in outsourcing and others do not.

When the decision is not dictated by business practices, there are many additional areas to consider.

Budget Type

For large corporations there are often two types of budget allocations: hard dollars and soft. Hard dollars are those that the company pays to its suppliers. Soft dollars are the funds transferred between departments. Obviously, contracting work with a service bureau requires hard dollars and internal operations use soft dollar transfers. While it may appear to be nonsensical, corporations may make this distinction, which could influence the implementation decision.

Delivery

With a service bureau, unless it is local, delivery time will include shipping. While most rapid prototypes will be shipped via overnight courier, this additional time can add one working day to the delivery time. With an internal operation that supplies prototypes to employees in the same facility, the prototypes can be hand delivered the moment they are complete. However, this assumes that both operations work with the same efficiency and speed. If an internal operation is not properly staffed or setup to run rapid prototyping machines around the clock, a remote service bureau may be able to deliver faster. Most service bureaus operate the machines 24 hours a day, including weekends and holidays. Additionally, the service bureau must stay current with the latest technology and materials, possibly making its technology faster.

The final consideration is that many service bureaus operate multiple machines. When compared to a single-system, in-house operation, the service bureau has the advantage of accommodating unexpected problems—system failure and bad prototype builds—and concurrently processing a large number of parts.

Breadth of Selection

As indicated previously, most service bureaus offer multiple technologies and multiple materials. Since no technology or material is perfect for every application, the service bureau offers choice and flexibility in applying the right solution to the application. For those organizations that implement one system with one or two materials, the internal users may at times need different properties and capabilities. The service, as an extension of the internal operations, becomes a cost-effective source of alternative technology and material solutions.

Other areas to consider reflect the changing business environment of the service bureau. To compete in an aggressive market, the service bureaus offer capabilities not often found within an organization. For example, some service bureaus offer instantaneous, online quoting of rapid prototypes. With this service, there is never a delay in receiving a quotation.

Realizing that prototyping is just one small component of the design and manufacturing process, some bureaus offer a single source, single point-of-contact solution for the entire product development and manufacturing process. From design to prototyping to prototype tooling to manufacturing, one source and one purchase order can get the job done. With so few internal resources, so little time, and so much to do, the full-service approach is often the most efficient and effective.

Finally, the service bureau can be an excellent source for advice and consultation. These companies have a wealth of information, not only on rapid prototyping but also on other prototyping, tooling, and manufacturing processes.

Consultation with the service bureau can ensure that the most effective and least costly solution is employed.

Safeguarding Proprietary Data

Disclosure of proprietary design data and product plans can be devastating if it falls into the hands of the competition. There is tremendous sensitivity to safeguarding this data. Obviously, when subcontracting rapid prototyping services, this valuable design data is beyond the direct control and protection measures of the corporation. For some, the sensitivity is so great that product design data cannot leave the facility without management approval.

While encryption and other techniques create an obstacle to the illicit acquisition of proprietary information, it is still possible to gain access to this data. Additionally, once the data is entrusted to the vendor, the organization has little control over what happens with it. The only real measure of protection is a legal device known as the nondisclosure agreement (NDA). The NDA is a binding agreement that makes the vendor liable for improper dissemination of confidential information. While every service bureau is willing to sign an NDA, most organizations realize it is primarily a preemptive measure that is unlikely to counteract the real impact of the disclosure of new product intents and designs.

Dual Mode Operations

As users of rapid prototyping have learned, it is unlikely that internal rapid prototyping capabilities will accommodate all of the internal demands. This may be due to capacity, lead time, material properties, or the technology capabilities. As many have found, it is wise to allocate some of the rapid prototyping budget to the purchase of prototypes from an outside source.

There are three reasons for an outsourcing strategy. First, it is not economically sound to have the available capacity to meet peak demands. Doing so means that on most days the system will be idle or underutilized. In this case, the service bureau provides capacity when demand outstrips supply. Second, for some rapid prototyping systems, carrying multiple materials can be expensive, and material conversion can be time consuming. Rather than bearing the expense of carrying all possible materials and impacting efficiency and productivity with downtime for conversion, parts are outsourced when desired material properties cannot be satisfied with the in-house material inventory. Third, it is unlikely that one technology can address all applications. It is best to implement a technology that addresses the majority of the demands while outsourcing the balance of work to service bureaus possessing the desired alternative technologies.

CONCLUSION

Obviously, there is much to consider when justifying rapid prototyping and selecting the best option for prototype development. With the extent of information offered, it may seem overwhelming and too time consuming to pursue. But do not let the amount of information lead to a conclusion that rapid prototyping is not a good solution. On the contrary, it is a powerful solution for most organizations, and it is one that must be considered in the quest to remain competitive.

REFERENCE

Wohlers, Terry and Grimm, Todd. 2002. "The Real Cost of RP." *Time-Compression Technologies*, March/April: 47–51.

CHAPTER 9

Evaluation and Implementation

THE CHALLENGE

The range of rapid prototyping technologies is quite broad, and with new developments, the range of options expands each year. There is diversity in the types of processes and the quality of the output. A broad spectrum of system prices and sizes exists. Wide ranges of material classes and material properties are available. With this variety and diversity, evaluation and selection of a rapid prototyping system can be daunting and difficult.

A lack of publicly available information, especially in the area of system limitations, hinders the evaluation process. Often, new owners of rapid prototyping technology learn of their system's limitations and deficiencies after the implementation. There are few resources and publications that discuss the weaknesses of the technologies. In the drive to position their technology for all rapid prototyping applications, some vendors may create unrealistic expectations in the marketplace. Without the insight of an experienced user or a thorough evaluation, the limitations may be unrecognized until after a system is purchased.

With dozens of processes and technologies from which to choose, selecting the right rapid prototyping system requires a thorough evaluation that builds upon a clear understanding of the corporate and departmental needs. As with other manufacturing tools, each system will have its strengths and weaknesses. There are no perfect systems for all users across all industries. A successful selection matches the primary needs for the majority of applications with the capabilities and constraints of the rapid proto-typing technology. The selection process must begin with a clear understanding of what is to be achieved, the gains, and the critical qualities and features required to deliver these results.

With a clear definition of requirements and a thorough evaluation, a suitable system can be selected. As many rapid prototyping owners have shown, the right technology can yield payback periods of less than a year. On the other hand, some rapid prototyping users have gone years without breaking even. The difference between these two scenarios could be either luck or diligence. Rather than leaving the success of the rapid prototyping operation to luck, potential users should do a diligent job of evaluation to ensure a successful rapid prototyping operation.

DEFINING REALISTIC NEEDS

The starting point of a rapid prototyping evaluation is the definition of system requirements. These specifications will result from a clear understanding of the intended applications and resulting requirements. However, it is also imperative to have realistic requirements for the application.

Inexperienced engineers may specify artificially tight tolerances on the parts they design. Likewise they may specify unreasonably loose tolerances that invite fit and function problems. Without experience and awareness of the impact of tolerance specifications on time and cost, the engineer may define inappropriate specifications for the application. When specifying rapid prototyping require-ments, a similar scenario is possible. Defining unrealistic requirements for a rapid prototyping operation can lead to excessive expense or poor performance.

To define realistic needs, an evaluation team should be assembled. As was described in Chapter 5, many disciplines within the organization use rapid prototyping. While the primary application is often design engineering, manufacturing, sales, marketing, and operations also may have applications for rapid prototyping. The first step in a successful evaluation is to define the disciplines that will be the primary users of the technology. From these disciplines, select representatives from each department for the evaluation team.

When building the team, it is also important to recognize that within each discipline, there may be multiple divisions and departments, and each may have their own needs. For example, companies with a diverse product offering will have designers dedicated to each product line, and each product type may have different needs. For example, one product line may comprise mostly metal castings, another injection-molded plastic parts, and a third may have a high concentration of sheet-metal stampings. This variety of applications and requirements should be included in the selection process. Therefore, the team may have multiple representatives from each discipline.

With a strong team in place, there should be an equally strong team leader. This individual should be comfortable and confident in making decisions that may not satisfy the whole team but are critical to the success of the primary users.

The next step is to define all of the applications of the technology, by department, and narrow the list of applications to those that will be the primary focus of rapid prototyping. Within design engineering, the choice of applications may be between concept models, form/fit models, functional models, and patterns. For manufacturing, the options may include tooling—rapid, bridge, short-run production, or production. For sales and marketing, the consideration may be for internal presentations, client proposals, trade-show prototypes, or advertising photo shoots. For many companies, all of these applications may be desirable. However, for the evaluation, the list of applications should be limited to only those that will have frequent use of the technology and the highest return on investment.

When defining the applications, the team should look toward the future, in terms of future products of the company, future applications, and future developments in rapid prototyping. To select a system with only present-day requirements may limit the usefulness and useful life of the rapid prototyping system.

After all this work, the team may be anxious to start the evaluation and get to the implementation. But that would be premature. The next step is to justify the acquisition of the technology and determine the level of investment that can be supported by the budget. From the list of primary applications, define the benefits the rapid prototyping technology will deliver and determine whether a system purchase can be justified. With consideration of budget constraints and the financial justification, the evaluation can be focused on those systems with an appropriate cost and the ability to deliver the desired results.

Because system capabilities and prices vary so dramatically, it can be beneficial to first select the class of technology that will be evaluated, which is often dictated by the available budget. If an evaluation includes all system classes—3D printers, enterprise prototyping centers, direct digital tooling devices, and direct digital manufacturing systems—the effort put into the definition of primary needs may be forgotten when the advantages of a higher-end system create new, previously undefined needs. Evaluating all classes simultaneously would be much like evaluating economy and luxury automobiles. While the economy car may meet basic requirements, the lure of the qualities and accouterments of the luxury vehicle may confuse the buying decision.

EVALUATING SYSTEMS

When evaluating rapid prototyping systems, there are two major components to investigate: materials and systems. Within these, it is important to evaluate the deliverable properties of the prototype, operational overhead, acquisition investment, and ongoing expense.

For many applications, the mechanical, electrical, and thermal properties of the prototype are critical to the application. Obviously, if the intended application is direct rapid tooling, the material will likely be metal or an equally robust alternative. Less obvious is the impact of material properties on applications like concept models or form and fit prototypes. While strength may not seem to be critical to these applications, many users find that good mechanical properties are still desirable. As rapid prototyping users have often voiced, material properties frequently outweigh accuracy, speed, and cost when it comes to a successful application.

With this in mind, it is often best to narrow the evaluation to only those technologies that can deliver suitable material properties. With this one decision, the number of systems for evaluation can be reduced to a manageable number.

Materials

Properties

The starting point of the evaluation is to reference the material properties data sheets. These will offer information on the strength, durability, flexibility, and heat resistance of the material. In some cases, the vendor may offer information on electrical conductivity, thermal conductivity, and

other properties. Since certain materials are generally positioned for specific prototype applications, the data sheets
will not be as thorough as those for production plastics and
metals. Material testing is expensive, and with the volume
of material sold annually, the cost of additional tests can be
prohibitive.

Another consideration is that the results presented in
material specification sheets should be viewed as a gener-
alized range of properties and reviewed relative to other
rapid prototyping materials. While material vendors adhere
to the International Organization for Standardization (ISO)
specifications for a given test, there is latitude in the testing
standards that can affect the results. A tensile test dictates a
specimen of a given shape, thickness, and temperature, but
these tests do not specify how the specimen is developed or
the time frame in which the testing may be completed. Variables such as the orientation of the layers to the direction of
pull or impact, elapsed time between building and testing,
and build parameters used to grow the specimen can
significantly alter the results. With this flexibility, vendors
may seek to optimize the parameters to deliver the most
beneficial results. Since it is unlikely that system operation
will have parameters identical to the testing standards, and
since the prototype may be in use for weeks, individual
results will vary. For this reason, do not accept the material
properties as a guarantee but rather as a realistic generalization of what is possible.

Speed

For the uninitiated, the hardware vendors claims of system
speed may be accepted as fact. For those who have used
rapid prototyping, there is awareness that prototype
construction speed is dependent upon a combination of the
hardware, prototype geometry, and selected material.

Different materials often have different build speeds. It is
reasonable to expect a speed variance of 30% or more for a

prototype built on the same system but in a different material. For example, in stereolithography the two critical variables that impact speed are critical exposure (E_c) and penetration depth (D_p). These two variables, which are material dependent, define the rate at which the laser can travel while curing a layer of the prototype to the desired layer thickness. Even the same class of material may have different rates of processing. For selective laser sintering, there is a difference between the draw speed of DuraForm® PA and DuraForm GF, even though both use the same base material. Since speed affects efficiency, capacity, and throughput, it is important to investigate this aspect of materials. The evaluation can be achieved with a vendor-built sample part, a service bureau project, or a discussion with current users of the material.

Cost

While the purchase price of materials for a given technology is generally competitive, those for different technologies can have tremendous variance.

To evaluate the true material cost requires a little math. Vendors sell rapid prototyping materials by the kilogram, pound, and spool. But, weight is not appropriate for a head-to-head comparison. For an accurate cost comparison, the material expense is based on a volumetric price (cost per in.^{3} [cm^3]). For example, the starch-based material for powder-binder printing is less dense than stereolithography materials. Therefore, the stereolithography prototype would weigh more than the binder-jetted part, but the volume of material would be the same (if prototype porosity is ignored). For an accurate cost comparison, acquire the material density to determine its price per volume.

As prototypes are built, there is invariably waste. Some systems have less than others, but there will always be some percentage of material lost with every build. Some of the waste occurs on the bench technician's table when supports are removed or excess powder is dusted away. Some systems, for example selective laser sintering, have even greater waste. In selective laser sintering, many of the powders require that a high percentage of new material be added to that which has already been processed. With typical usage, this blend of new and used material may lead to a stockpile of used material that is never reclaimed.

For those systems replenished with new material as builds are completed, there is the possibility that the material becomes unusable. Stereolithography, for example, imparts small amounts of UV energy to the uncured resin. As the resin ages, the cumulative UV exposure may change the characteristics of the raw material, making it unusable.

Safety and Environment

In the earlier days of stereolithography, when acrylate-based resins were common, those exposed to the resin, either through respiration or skin contact, were at risk of sensitization. Once sensitized, exposure to the resin could result in rashes, hives, and swelling. While the newer stereolithography resins do not appear to have the same effect, safety and protection are still issues to consider. Even innocuous materials like powdered polyamide or starch can be a cause for concern. In a fine powder, these materials are easily airborne, posing respiratory hazards.

Many rapid prototyping materials cannot be disposed of in the trash or down the drain. Special handling and disposal regulations may apply, and these will vary from state to state and company to company. Evaluate these areas to understand the total material cost for the rapid prototyping operation.

Other Considerations

It can be tempting to select a new material that has exciting properties and promises tremendous opportunities.

Selecting such a material can have powerful rewards, but an equal amount of risk is accepted. Operation in a research and development lab can be quite different from that in a production environment. Commonly, the true behavior of a material is not realized until a significant number of users have processed it in their own operations. In the past, promising materials have failed when they reached the real-world environment of rapid prototyping. For this reason, it is often best for a new rapid prototyping operation to rely on an established and proven material.

The final material consideration is the high cost for material conversions in some technologies. Some users operate their systems with only one material to avoid the expense and downtime of a material swap. Thus, if a variety of material properties is required, it may be best to select a system that offers easy and inexpensive options for material conversion.

Systems

While it will be fairly easy to discover the strengths of each technology, determining the limitations and operational constraints may prove difficult. Finding this information will require investigation; talking with users, attending conferences, and possibly seeking outside assistance.

It is important to reiterate that the hardware review must be conducted with respect to the identified applications and needs. To begin the selection, first define all of the potential applications of the technology. For example, will the system be applied to conceptualization, form, fit and function analysis, pattern generation, tool design and creation, or rapid manufacturing? The second step is to take these applications and list the requirements for each. The considerations could include accuracy, material properties, or physical size. With the list of requirements, the evaluation of the available technologies can begin.

Physical Properties

As emphasized in the material evaluation, rapid proto-typing is a system with output that is dependent on the hardware and selected materials. In reviewing the tech-nologies, both must be concurrently evaluated to under-stand the available physical properties that can be deliv-ered. Additionally, results can vary with the skill level of the operator.

Dimensional accuracy. Each rapid prototyping system will deliver varying degrees of dimensional accuracy, which are influenced by the resolution and precision of the building process and the layer thicknesses available. Yet, many other variables affect the deliverable accuracy of prototypes.

System manufacturers offer claims of accuracy for their systems, but these should be treated, in general, as a best case deliverable. For a system that advertises ±0.005 in. (±0.13 mm), it is likely that some of the dimensions fall within this range while others deviate significantly. Even if variables like material shrinkage, operator experience, and temperature could be removed, the dimensional results are still likely to deviate from the vendor specification as the size of the prototype grows. Most of the systems deliverable accuracy is a function of the size of the prototype; larger prototypes have greater dimensional variance.

Typically, smaller beam size, droplet size, filament size, or powder particle size will yield tighter dimensional accu-racy. Likewise, thinner layer thicknesses will further improve accuracy. For example, in the stereolithography system line-up, the Viper si2™ is recognized as a high-accuracy device that replicates small features. This system offers a beam size of 0.003 in. (0.76 mm) and layer thick-nesses as small as 0.002 in. (0.05 mm). However, when building prototypes with this resolution, time increases. Most often, users of the Viper si2 will reserve the use of the small laser spot size and thinnest layers for only the most demanding applications. Thus, when evaluating accuracy for the Viper si2 or any other system, it is important to consider dimensional tolerance in light of acceptable build times.

As mentioned previously, for most rapid prototypes, the deliverable accuracy is a function of the technology and the model making staff. For those prototypes that require any degree of finishing, the talent of the bench technician ultimately defines the accuracy. As a result, the amount of finishing and its impact on accuracy for a given technology in a specified application must be considered in the evaluation of accuracy.

Surface finish. The quality of the finish of a prototype is important in terms of cosmetic appeal and impact on the intended application. While every rapid prototype can be finished to any desired level, the amount of labor to achieve the finish is a consideration. During the evaluation, it is important to review finishing requirements in light of reasonable and acceptable surface finish.

Surface finish can go well beyond cosmetic appeal. In applications such as functional testing and pattern generation, the quality of the finish may have significant impact on the results of the project. For example, if a rapid prototype is to be used as a pattern to create a mold or tool, a rough surface finish will affect the quality of the molded part and the quality of the tool. For rigid tooling, a poor surface finish may prevent ejection of the molded part if the surface defects form areas in the tool that serve as mechanical locks. Flow testing offers another example. If the surface of the prototype is rougher than that of the final product, the flow testing results may be impaired. For both of these examples, applying labor in the area of part benching can resolve the problems. However, if the surface is inaccessible, it cannot be finished and will have to suffice.

Due to the stair stepping inherent in rapid prototyping, there is always the need for benching when the prototype is applied to higher-level applications. The key question is whether the finish is acceptable for the application or whether the amount of labor is acceptable to make the rapid prototype surface finish appropriate for the application.

Size. The working envelope of the rapid prototyping system will dictate the maximum size of the rapid prototypes. Additionally, the build envelope will impact throughput and operational efficiency.

Obviously, the capacity of the system should accommodate the typical range of prototype sizes required of the users. Although prototypes can be built in sections and later bonded, this creates additional work and expense, and will delay the delivery of the prototype.

Since rapid prototyping systems may gain efficiencies when constructing multiple parts in one build, the system's build envelope should accommodate the expected mix and quantity of prototypes. If the system size is too small, a backlog of prototype work could develop, which would delay deliveries. If the system size is too large, capacity may be wasted and efficiencies lost. Similarly, if targeting maximum efficiency, an oversized system may extend prototype deliveries as builds are delayed while waiting for enough parts to fill the envelope. For those systems that produce waste material with each build, an oversized system may also lead to excessive material costs.

During the evaluation, the actual build envelope needs to be determined. Some system vendors advertise the size of the build chamber, not the true size of the working area. Therefore, investigate the usable build envelope that most users consider practical.

Repeatability. The final physical aspect is the repeatability of each of the previously discussed factors. Since systems can be sensitive to temperature, humidity, age of material,

material reclamation, power levels, and maintenance, prototypes produced on different systems of the same model often produce varying results. With additional considerations, such as part orientation and placement, results can also vary for additional copies of the same part produced on the same machine.

Operational Criteria

Operationally, systems can vary to a great degree. While some systems can be installed and operated with little training and site modification, others require training, staffing, and facility modifications. Likewise, throughput and build time can vary greatly. These operational criteria are critical in an evaluation of rapid prototyping systems. They will affect not only the capabilities of the rapid proto-typing lab but the investment and throughput. Combined, these considerations can contribute to the long-term success or future failure of the in-house rapid prototyping operation.

Build time. Since rapid prototyping devices are often purchased for their speed, the most obvious criterion is time. Time can be viewed in terms of build time for a proto-type, total cycle time for a prototype, or throughput time. Comparing systems in terms of build speed for a sample part is myopic and often does not offer the whole picture. While one system's build speed may be four times faster than another, the total time from data receipt to prototype delivery may be nearly identical.

Build time is often cited by vendors as cubic inches (cubic centimeters) per hour or vertical inches (centime-ters) per hour. This can be misleading and should be treated as only an approximation of the relative build speed. As stated previously, the material applied to the job can dramatically impact time. Additionally, other factors such as support structures, geometry type, build style, and part design may alter the build time. For example, in stereo-lithography, a variation in build style and user-defined

251

parameters can alter build time by 50% or more. Additionally, due to the additional laser energy imparted into support structures and boundary layers, an increase in the number of supports or the surface area of a part will increase build time. In this instance two parts, both with a volume of 1 in.^{3} (16 cm^3) and 1 in. (25 mm) tall, would differ significantly if one were a solid cube and the other a pill-shaped, hollow part.

Since layer thickness has a significant impact on build time, analysis of competitive systems should be completed with comparable (or equal) layer thicknesses most common for everyday use. For every system, there is a minimum time per layer, and this means that as the layer thickness decreases the build time increases.

A final consideration in determining build time is the requirement for any pre-build warm up or post-build cool down. This time requirement is not frequently discussed, and some users are surprised to find that a one-hour prototype actually takes three hours with the addition of warm up and cool down. The rapid prototyping system may require preparation prior to building (purging, material addition, etc.), a warm up period prior to building, and a cool down period after the build is complete. These numbers are rarely included in vendor speed claims, but they are factors in the total build time. In some cases, they can be significant factors, so it is important to qualify this element of time in any evaluation.

Total lead time. Total lead time should consider all of the work (once again with the primary application in mind) after the prototype is delivered from the machine. Different systems require varying degrees and types of cleaning and benching, so it may be found that a rapid system actually produces prototypes slower in terms of total cycle time. Also, some systems offer labor-less cleaning that may take more time but require little, if any, staffing. If a corporation operates leanly, the manual efforts could become the bottleneck and constrict the throughput while increasing total cycle time.

Throughput should be considered in two ways: the throughput capabilities of the machine and the throughput capabilities of the shop. System throughput consists of the combination of build time and system capacity. A fast system with a small build envelope may have lesser throughput than a slightly slower system with larger capacity. In this regard, some estimate of total demand may assist in determining the ideal capacity utilization, production rate, and throughput.

As discussed previously, rapid prototyping becomes a manual, labor-intensive operation once the build is complete. Consider the time to clean and bench a prototype and the available staff. If the staff is too lean for the demands, throughput and delivery time will suffer.

Staffing. Labor can be a significant component of the cost of rapid prototyping operations. It will vary by system, application, and operation.

Nearly all rapid prototyping systems can be operated by individual designers or engineers, which eliminates the need for dedicated staff. However, this does not mean that it is the right way to run the equipment. For those systems requiring skill, ingenuity, and experience, the best alternative is to hire personnel who are dedicated to the rapid prototyping operation. When evaluating the staffing requirements, it is also important to gain a sense of the time involved in routine system maintenance, repairs, oversight, and management.

Unattended operations are possible for some systems. However, for others it may be wiser to have an operator who can observe the device while it is in operation. For those organizations that plan to run the rapid prototyping system 24 hours a day, seven days a week, a support person will be important. To gain maximum operational efficiency, operators will need to be on hand to unload a completed build and start the next one.

Finally, to gain the full advantage of the rapid prototyping system and solve problems that develop, a dedicated staff person is needed. Rapid prototyping continues to be a bit of an art, especially for challenging prototypes and applications. For those systems that have yet to become a push-button operation, having an individual whose primary role is the operation of the system is often an important factor in the success of the technology.

Equipment and facilities. As discussed in Chapter 8, facility and equipment requirements are highly dependent on the technology selected and the intended application of the prototypes.

Since the cost of facility modifications and supplementary equipment can approach, and even exceed, the cost of the rapid prototyping system, it is imperative that these requirements are determined during the evaluation. The minimum requirements and recommended configuration can be acquired from the system vendor. However, the investigation may need to include current system users to determine the optimal configuration.

Finally, consideration should be given to corporate standards and governmental regulations. Fire, health, and safety standards may require further facility modification or additional equipment.

Software. Software evaluation can be simple and straightforward when compared to the material and hardware evaluation. The key software components for rapid prototyping are 3D CAD and rapid prototyping preprocessing tools.

Without an existing 3D CAD system implementation, it is unwise to consider a rapid prototyping system. The process of selection, implementation, and transition to 3D CAD technology is a major undertaking. Thus, software selection is simplified because it must be in place prior to a rapid prototyping evaluation.

While there are a number of software preprocessing tools to consider, it is recommended that this evaluation be delayed until after the successful implementation of the rapid prototyping system. Each system offers the fundamental tools to prepare STL and build files. Thus, a functioning rapid prototyping operation is possible without additional software. After a period of operation, the true needs for preprocessing software will be discovered. If it is found that the operation would benefit from third-party applications, an evaluation can then be conducted.

Conclusion

Obviously, with so much dependent on the selected material and the build parameters used to construct the prototype, the evaluation of any rapid prototyping technology will require information from multiple sources. To rely on the results of a single sample prototype as being indicative of deliverable results would be ill advised. A thorough and accurate evaluation demands that information be secured from multiple sources and that this information relates to the combination of technology, hardware, material, and construction parameters being considered.

There are many aspects of rapid prototyping systems that are not obvious. There may be hidden costs, concealed limitations, and unseen obstacles. Yet, these factors will be hidden only if they are not investigated. A thorough evaluation that obtains information from vendors and users will lead to a successful operation that matches the requirements established in the planning phase.

IMPLEMENTATION

The implementation of a rapid prototyping system comprises three key areas: front-end operations, rapid prototyping, and back-end operations. While some system implementations can be easy and simple, many are more complex and involved, requiring facility modification, training, and new processes and equipment.

As with the evaluation, the implementation, especially back-end operations, should focus on the primary applications and needs that have been identified for a successful and beneficial operation. Additionally, since rapid prototyping must be fast, consideration of the process and work flow in the operation should be given before, during, and after the implementation of the technology.

The System Needs a Champion

Prior to discussing the three areas of implementation, an often ignored element of a successful implementation and operation needs to be brought to light: a champion. Every rapid prototyping operation needs a champion to promote the technology, oversee operations, and justify the department—or the technology—existence.

What is a champion? A champion is an employee with some rank within the organization and some decision-making authority. But what makes a champion different from a supervisor or manager is that the champion actually champions the cause. The champion is part cheerleader, part politician, part customer-service representative, part evangelist, and an enthusiastic supporter of the technology. The champion is a believer with a personal stake in the success of the technology.

A champion is willing to present a justification for system purchase or additional staffing to those who make the final decision. The champion is willing to persist until a satisfactory decision is made. A champion promotes the technology and its application throughout the organization to ensure all are aware of its capabilities and benefits to speed internal adoption.

As operations commence, the champion makes sure the technology and its supporting staff deliver as promised. The champion is always aware of internal customer satisfaction. The champion consistently monitors and reports the value of the rapid prototyping operation to upper management. Without documented gain, it is easy to overlook the intangible value of quickly delivering prototypes when an organization is focused on the impact to the bottom line.

Front-end Systems

To deliver the speed and responsiveness expected of rapid prototyping, the implementation of front-end systems must address process and computing needs. For the front-end systems, the areas addressed relate to data processing, including receipt, management, and preparation of data for the rapid prototyping builds.

Rapid prototyping is fast paced and subject to frequent change. Therefore, a process to coordinate, manage, and schedule the operation is vital. As with shop-floor operations, a system, either paper or computer based, is required to document the jobs in queue, schedule new orders, and track the status of work in progress. For small operations, a spreadsheet may suffice, but for larger operations with higher prototype volumes, a database application may be most appropriate. Independent of the method chosen, the system must allow flexibility to accommodate a rapidly changing work schedule. In a busy rapid prototyping operation, the schedule will change hourly.

The process should include a system for the management of STL files and CAD data. Multiple iterations of a part will pass through the rapid prototyping operation. In some cases, users find the data changes between the time an order is submitted and the time the build commences. To avoid construction of earlier revisions, a system for the archival of files should be created.

Beyond the process aspect of the front-end, the installation of hardware and software is relatively straightforward. For any organization that has had past information technology (IT) system implementations, most of the elements are common to, and can build from, past projects. Important components will include a pathway for data transmission and computers and software for file preparation.

While CAD and STL files can be relatively small, most will range from 0.5–50 MB. There is the possibility, which is dependent on the types of products produced, that these files could swell to 100 MB or more. With files of this size, it is important to have a network that can rapidly transfer them. For single-facility operations, this may include only a local area network with fast transmission speeds or the ability to write, deliver, and read CD-ROMs. For multi-location operations, a wide area network with fast data transmission is required. Options include DSL lines, T1 lines, and FTP servers. Establishing the ability to transfer data rapidly is important to the efficiency of the operation and its ability to deliver quickly.

To process the data, software applications are installed and operators trained. With the proliferation of Windows®-based software applications, this step of the implementation is straightforward. There is a likelihood users may be able to operate the software with little or no training. The software applications will come from the vendor of the rapid prototyping system. In addition, third-party software for support generation, file repair, and data manipulation may be selected.

Rapid Prototyping Systems

Implementing the hardware for a rapid prototyping operation requires some advanced planning. Prior to the delivery of the system, facility modifications are often required. Most rapid prototyping systems are best suited to a controlled lab environment, not the shop floor or an office area. In constructing the lab, considerations include HVAC, isolation of airborne contaminants, and electricity. For some systems, supply lines for gases or water also may be required. In the lab, allot space for material inventory, tools, and supporting equipment.

System vendors can offer guidance on the setup of front-end and back-end operations. Each will be able to offer all of the details of the implementation of the equipment and the requirements for facility modifications. For this phase, rely on the vendor's plans and information, and allow an appropriate amount of time and budget for completion of the facility.

Once the facility has been modified to the vendor's specifications, the hardware implementation can begin. Note that for some systems, this may take just hours, but for others it may take days. In general, the equipment is unpacked and placed, and in some cases, components assembled. When the system is fully installed, it will be calibrated and tuned to the vendor's specifications. Finally, a test part will be constructed and offered for the buyer's approval and acceptance of the system.

Staffing and Training

With the fast-paced and ever-changing schedule of a rapid prototyping operation, it is usually unwise to make system operation a subordinate task to an employee's primary duties. Except for those systems that require little training and no expertise to build a prototype, it is best to have an employee whose primary job is system operation and maintenance. Once the operators have been identified, they should be sent to vendor training programs.

For most systems, operation is not by rote with step-by-step procedure, but rather a process blending the operator's skill, creativity, and problem-solving abilities. To gain this experience, the operator may require some time to exercise the system and gain hands on experience.

Prior to the installation, practices and procedures should be created for system operation and material storage, handling, and disposal. For nonhazardous materials, the procedures may focus only on proper handling, disposal, or reclamation. For systems that use hazardous materials, employee safety procedures should be created and corporate policies and governmental regulations reviewed. Supporting safety equipment (gloves, respirators, lab coats, etc.) should be purchased in advance.

In all cases, the equipment vendor will be an important information source for facility, safety, and equipment requirements. This information will be offered in advance of system delivery so the facility is ready for the installation.

Back-end Operations

Post-build requirements of rapid prototyping systems vary greatly, especially with the variety of applications and associated post-processing demands. This yields dramatically different implementation needs that are highly dependent on many factors.

In general, there are two components to consider during cleaning and benching. Cleaning is the process of making the prototype suitable for consumption in its most basic form. In a sense, the cleaning process delivers an "as-is" model. Benching is the process of taking the prototype to a higher level of quality. The benching process is similar for any prototype or model to which a model-making team applies their skill.

Cleaning

While not every model requires benching, every rapid prototype will require some degree of cleaning. Considering the four systems reviewed in this book, this includes support removal, powder removal, resin stripping, post curing, and washing. In some instances, treating the surface of the prototype with an infiltrant is also considered to be part of the cleaning phase.

For each system, there will be a minimum specification for cleaning equipment, procedures, and supplies. However, beyond that, it is often individual preferences that dictate how the parts will be cleaned. While three of the systems prototypes could be cleaned in an office environment, this is not advisable. Odors and debris may be generated that may create hazards. Thus, the implementation should consider the allocation and design of an area suitable for cleaning the parts from the process.

The area created for cleaning should accommodate manual stripping of supports, manual removal of excess powder, and a washing station. In the case of stereolithography and fused deposition modeling (when Water-Works™ is used), solvent tanks also will be required. For stereolithography, a solvent strips the viscous resin from the part surface, and for fused deposition modeling the tank contains a solvent that dissolves the WaterWorks support structure. For stereolithography, one other piece of equipment is required, an ultraviolet (UV) oven. Stereolithography parts do not emerge from the machine fully cured, so a UV oven is required to further cure the resin in the prototype.

Optional, but recommended, cleaning operations for the powder-based systems require de-powdering stations, tools, and surface-treatment agents. These surface treatments wick into the pores of the prototype to either increase durability or provide a sandable surface.

Due to Occupational Safety and Health Administration (OSHA) and corporate and general sensibility requirements, cleaning operation implementation should include protective equipment for staff (respirators, safety glasses, gloves, etc.). In addition, there should be a plan for waste material handling and disposal. When solvents are used, it is most likely that the waste material cannot be disposed of down the drain. Additionally, reclamation and/or disposal of residual powder may need to be accommodated.

Obviously, the cleaning process requires staffing. For a single-system operation, the cleaning area is most often staffed by the rapid prototyping system operator. For larger operations, additional staff will be required. No matter who and how many, the implementation plan should include the identification of staffing needs and training for these individuals.

Benching

The considerations for benching operations are similar to those for any model shop environment. In fact, if a model shop exists, the implementation may require only the addition of a few pieces of specialized equipment.

Benching is the most labor-dependent operation in the rapid prototyping process. For every system, supplying a prototype, pattern, or tool with the desired level of finish will require some degree of benching. This process will require facility modification for workstations, solvent baths, debris isolation, and possibly paint booths. Additionally, an inventory of supplies and tools is necessary.

Benching of rapid prototypes requires many of the elements of a model shop. For those organizations with an existing model shop, this step of the implementation can be as simple as training the existing staff on the proper procedures for benching the rapid prototypes. For those without an existing model shop, this part of the implementation can take much time and effort. In fact, it is often best to build this capability to only the minimum level required to accommodate the desired applications. As time passes and experience is gained, the benching operation can be expanded as needed.

Having the necessary tools is important, but staffing is perhaps the most important aspect of benching. Once a rapid prototype is delivered to the shop, the bench technician controls the quality. It is no longer a function of the system. Additionally, many find that benching is the bottleneck operation, the one that throttles the speed of delivery. When staffing this operation, look for a unique personality type: a person who is always concerned with details but can do so under constant time pressures. Additionally, consider the existing skill level of candidates. While anyone can sand parts, not everyone can do this with the care, precision, and proper style of a model maker. Finally, it is often best not to staff this operation entirely with model makers. Instead, employ one or two model makers to supervise, train, and mentor the staff while handling the tougher jobs. The reason is simple: model makers are artisans skilled at crafting parts, and benching for rapid prototyping is most often a tedious, uncreative process.

Facility Design and Equipment

The facility design and equipment required will vary by application and slightly by the chosen technology. Yet, in general, the following should be considered and planned for in the implementation phase.

For the facility, there will be a requirement for workstations, power, lighting, compressed air, ventilation, and debris isolation. Each bench technician will need a work area of suitable size that is well lit. The facility should be large enough to accommodate the workstations and any supporting equipment. This equipment may include paint booths, power tools, and basic shop equipment.

Finally, as with the cleaning process, the work area should have proper ventilation and debris isolation. In some cases, the area should have reasonable HVAC and humidity control. For example, in the case of stereolithography, some materials are temperature and humidity sensitive. Prolonged exposure to either condition on the shop floor may lead to dimensional change or part deformation.

The equipment needed for benching will depend on the application and personal preferences. At a minimum, it will require sandpaper, sanding blocks, files, X-Acto® knives, and small picks/probes. To decrease time, power tools may be added, such as Dremel® tools, orbital sanders, and disk sanders. For advanced applications, the user may desire to add some shop tools such as a drill press, lathe, or mill. And finally, if beautifully painted prototypes will be the deliverable, paint sprayers and booths should be considered. While a can of spray paint may get the job done, the quality will not match that of a professional setup.

The benching operation will also require other chemical agents beyond paint and primer. Various adhesives and solvents may be employed.

Working around the dust and vapors of the shop floor, safety should not be forgotten. Procedures and requirements should be documented before the first part is processed. Safety equipment may include eye protection, respirators, dust masks, and gloves. Additionally, with some chemical agents there may be special storage and disposal requirements.

Benching operations generate waste and contaminants. Therefore, thought should be given to the disposal of wastes (some considered hazardous), safety, and isolation of airborne contaminants.

If the prototypes will not be hand delivered to an office around the corner, the final consideration is packing and packing supplies. After crafting the perfect prototype on time, nothing is more disappointing than to have it damaged in transit. Packing supplies may include boxes, bubble wrap, packing peanuts, packing foam, and tape.

CONCLUSION

A good plan for evaluation and implementation of rapid prototyping systems is the foundation for long-term success. A clear, realistic definition of the requirements of a rapid prototyping system is the starting point for a successful evaluation.

CHAPTER 10

Rapid Tooling and Rapid Manufacturing

Rapid prototyping has demonstrated its ability to reduce time and cost in the development cycle while improving product quality. With this success, the industry's attention has turned to downstream processes that promise an even greater impact on time and cost. These applications are rapid tooling and rapid manufacturing.

Early rapid tooling initiatives addressed the applications for prototype tooling. Today, rapid tooling can encompass prototype tooling, bridge-to-production tooling, and production tooling.

Close on the heels of rapid tooling developments, another effort arose: rapid manufacturing. Applying rapid prototyping technologies to the manufacturing process would eliminate the need for tooling of any kind. In doing so, cost and time could be slashed.

For many, it may be premature to consider either rapid tooling or rapid manufacturing. There are limitations in both that prevent widespread use. Yet, as developments unfold, both will be very powerful in the future.

DEFINITIONS

It is important to reiterate the definitions of rapid tooling and rapid manufacturing as they are used in this book. Almost as soon as the terms were coined, every conceivable process and technology jumped on the rapid tooling and rapid manufacturing bandwagon. Today, any technique, technology, or process that produces tooling or manufactured parts quickly is likely to be labeled as a "rapid" solution. While entirely accurate in describing the end result, without clarity of definition, it would be difficult to clearly and concisely discuss these two technologies.

The definitions are:

Rapid tooling: The production of tools, molds, or dies—directly or indirectly—from a rapid prototyping technology.

Rapid manufacturing: The production of end-use parts— directly or indirectly— from a rapid prototyping technology.

Additionally, these technologies are limited to those processes that deliver production intent materials. The reason for this additional constraint is that in rapid tooling applications, many refer to processes that produce components in materials that mimic the intended production material as rapid tooling or rapid manufacturing. For example, some organizations classify rubber molding for cast urethane parts as both a rapid tooling and rapid manufacturing application. While rubber molding quickly delivers molds and parts from a rapid prototyping pattern, the material is a thermoset, which at best mimics the material properties of an injection-molded thermoplastic resin.

LIMITED USE AND SUCCESS

With the powerful benefits of rapid tooling and rapid manufacturing, it would seem that a large percentage of companies would commonly use these applications for many products. It would also seem that any company that did not use these techniques would face competitive forces that threaten their existence. Some day, in the future, this will be true. But for now, both applications have found limited use and success.

As will be discussed in this chapter, rapid tooling and rapid manufacturing have limitations that constrain their use on a broad spectrum. These limitations come in many forms: accuracy, surface-finish, and tool life, to name a few. Additionally, the threat to conventional processes, like the CNC machining, has given impetus to improvements in time and cost in these competitive technologies. In doing so, the advantages of rapid tooling and rapid manufacturing have been diminished while the limitations remain relatively constant. For example, in the mid-1990s machined prototype tooling commonly took 4–12 weeks. Today, with advancements in software, cutting speed, and shop-floor processes, the lead time for a prototype tool is now in the range of 2–6 weeks, a 50% reduction.

Equipment manufacturers, material suppliers, universities, research organizations, and end-user companies are researching and developing advancements in rapid tooling and rapid manufacturing. Over the next 5–10 years, this research will yield advancements that make the application of these technologies more commonplace. As companies adopt these processes, they will fuel further research and development, leading to even greater application of rapid tooling and rapid manufacturing.

RAPID TOOLING

In the quest to reduce cost and time in the construction of prototype, short run, and production tooling, many have evaluated rapid prototyping technologies as a solution. While some have had great success with rapid tooling, its application has been limited, making it more of a niche application. As advances in machined tooling have driven out cost and time, rapid tooling's advantages have diminished, and its limitations remain unchanged.

There are two methods of achieving rapid tooling. Indirect methods use a pattern generated from a rapid prototyping device. The pattern is used to cast or form molds or tools in a variety of materials, including epoxy, kirksite (a low-melting-point alloy), aluminum, and metal alloy blends. Direct methods produce tools or tooling inserts from the rapid prototyping device. Materials for direct methods include many metal alloys, alloy blends, ceramics, composite materials, and even rapid prototyping plastics.

Indirect Processes

There are many methods of indirect rapid tooling. To illustrate the process, a few common techniques are described in the following sections.

Epoxy Tooling

A two-part, aluminum-filled epoxy resin is poured onto a pattern. The epoxy cures to create a rigid tool. Since epoxy in thin sections cannot withstand injection-molding pressures, metal inserts are often machined and incorporated into the tool during the casting process (see *Figure 10-1*).

Spray Metal Tooling

Spray metal is a commercial process where metal powder or wire is melted, atomized, and projected onto a surface at high speed. In commercial aerospace applications, many alloys are used, but in spray metal tooling, low-melt alloys are commonly applied, since the temperature and required spray velocity of other alloys would destroy the pattern.

For spray metal tooling, the metal alloy is sprayed directly onto a pattern to form a thin metal shell. This shell

FIGURE 10-1. Often generated with rapid prototyping, epoxy tools are cast from patterns and used for injection molding of many resins and engineered plastics. *(Courtesy Ralph S. Alberts Company, Inc.)*

is then backed with epoxy or a castable, low-melt-temperature alloy to form the tool. Since the process requires line of sight and accessibility of the spray to the part surface, some features, such as deep, narrow channels, require machined metal inserts.

Cast Kirksite Tooling

Plaster mold casting is commonly used to create aluminum, zinc, or magnesium parts. However, it can also be applied to tooling applications. From a pattern, an intermediate tool is created in rubber or epoxy. The intermediate tool is then used to make a dispensable plaster mold into which metal is cast. Often, kirksite, a low-melt alloy, is the cast metal used for the tool. An example of a cast kirksite tool is shown in *Figure 10-2*.

FIGURE 10-2. Cast kirksite tooling is constructed using the plaster-mold-casting process. Shown here are a core and cavity with the original stereolithography patterns resting on the cavity. *(Courtesy Armstrong Mold)*

3D Keltool®

A mixture of metal powder and binder are packed against a pattern and allowed to cure. The "green state" tool is then fired in a furnace to burn away the binder, sinter the metal, and infiltrate the remaining cavities with a metal alloy. *Figure 10-3* shows a selection of Keltool cores and cavities.

Limitations

While indirect processes can be very fast in the creation of tooling, the dependency of time and cost on part size and complexity can easily push delivery to a month or more. With the speed gains in machined tooling over the past

FIGURE 10-3. The ability of 3D Systems' 3D Keltool® to capture fine detail is demonstrated in these photos of various tooling cores and cavities. *(Courtesy General Pattern Company)*

decade, for many applications, indirect methods may have a minor time advantage or no advantage at all.

Other considerations of indirect tooling include low tool life, typically 50–1,000 pieces, and limitations of the resins that can be injection molded. While many common thermoplastics can be molded, engineered thermoplastics or filled resins often exceed the operating parameters of an indirect rapid tool. One exception in both cases is the 3D Keltool process, which is capable of molding large quantities of most thermoplastics.

The indirect processes require an intermediate step, the conversion of positive pattern to negative mold, which introduces additional tolerance deviation. While patterns can be adjusted to account for shrinkage during the formation of the tool, it is difficult to predict on a feature-by-feature basis. The end result is that indirect methods are unlikely to deliver the tolerance expected of a machined injection mold. Additionally, the rapid prototyping pattern may be prone to distortion under the forces of casting, forming, or spraying the tooling material. This directly translates to additional inaccuracies.

The final considerations include long cycle times and higher part prices. With the exception of 3D Keltool, the heat transfer properties of an indirect rapid tool require longer cycle times on the injection molding press. Since these tools retain heat longer than aluminum or steel, the cycle time is extended to allow the tool to cool to operating temperatures. To deliver fast, cost-effective solutions, most indirect rapid tools are designed for manual operation, as opposed to the automated processes of a short-run or production tool. This additional labor adds to the cost of the molded part. Typically, part prices would be approximately 10 times that of a production-molded part.

Direct Processes

Direct rapid tooling encompasses many processes. A few common techniques are described in the following sections.

Sintering

The selective laser sintering process, like others, offers the ability to sinter metals that are suitable for tools or tooling inserts. Since the laser's wattage is relatively low, the metal is not sintered in most cases; rather, a binder that coats the metal powder is what is fused. After construction in the rapid prototyping system, the "green" part is fired in a furnace where the binder is burned off and another alloy is infiltrated into the pores of the tool (see *Figure 10-4*). Commonly, the base metal will be a strong, hard alloy such as stainless steel, and the infiltrant will be a softer, lower-melting-temperature material like bronze or copper.

Powder-binder Printing

In powder-binder printing, a binder is deposited onto a bed of metal, ceramic, or silica powder. When producing metal

FIGURE 10-4. With selective laser sintering and LaserForm™ ST-100 from 3D Systems, injection molds are produced directly from a rapid prototyping system. The resulting tool is 60% stainless steel and 40% bronze. *(Courtesy Bastech, Inc.)*

tooling, the tool is fired in a furnace to burn off the binder and infiltrate the pores with another alloy. For metal tooling, Extrude Hone offers the ProMetal® system, which is based on the MIT 3DP technology. A lost-foam tool produced on a ProMetal system is illustrated in *Figure 10-5*. For sandcasting applications, Z Corporation offers the ZCast™ process pictured in *Figure 10-6*.

Metal Deposition

Over the past several years, several technologies have emerged that directly deposit metal alloys on an additive basis. Unlike selective laser sintering, the metal deposition process melts and deposits, hard, strong metal alloys on an additive basis. The result is often a fully dense tool

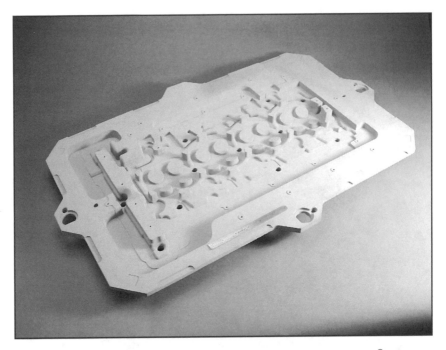

FIGURE 10-5. A large, lost-foam tool created on a ProMetal® system, which jets binder onto a bed of metal powder. Note the 6-in. (152-mm) scale in the foreground. *(Courtesy Extrude Hone)*

constructed of durable materials commonly used for machined tooling. *Figure 10-7* shows one such system, a POM DMD™ 3300, in operation. POM calls their process direct metal deposition (DMD).

Metal Lamination

While there are a variety of strategies to construct tooling through the lamination of metal sheet, one process stands out, ultrasonic consolidation. This process, from Solidica, Inc., ultrasonically welds aluminum tape. To create the profile of the tool, the welded layer sheets are milled. In effect, this process incorporates both additive and subtractive methods (see *Figure 10-8*).

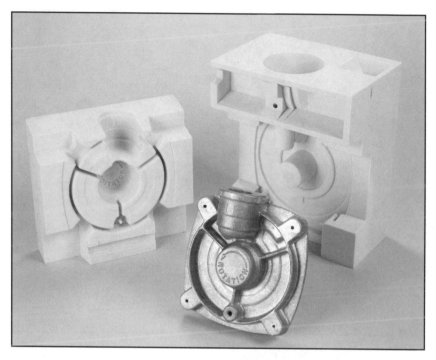

FIGURE 10-6. The ZCast™ process produces molds in a plaster-ceramic composite material for sandcasting applications. Metal is cast into the tool, and when solidified, the tool is destroyed to yield the metal casting. *(Courtesy Z Corporation)*

Limitations

While direct methods of rapid tooling overcome many of the limitations of indirect tooling, this in turn creates new barriers. The primary issues involve time, tool size, accuracy, and surface finish.

Since most rapid prototyping systems' build time is a function of the volume of material in the part (or tool) and the height of the part, construction of tooling inserts can take a significant amount of time. For even reasonably sized inserts of less than 8 × 8 × 8 in. (203 × 203 × 203 mm), construction time can take more than a day. When tool design, build time, and secondary processes are consid-

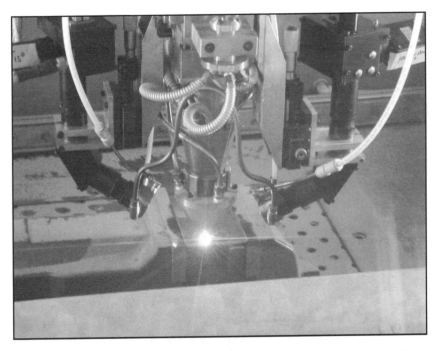

FIGURE 10-7. Direct rapid tooling solutions include processes such as direct metal deposition (DMD™) from POM Group. Pictured here is the DMD 3300. *(Courtesy University of Louisville Rapid Prototyping Center)*

ered, the delivery of a direct rapid tool is commonly in the 1–2-week time frame, not 1–2 days as with prototyping.

With the build envelopes available in rapid prototyping systems, it is most likely that tooling inserts are created. For injection molding, tooling is often two to four times the size of the part to be molded. Maximum build envelopes that approach 3 × 2 × 2 ft (0.9 × 0.6 × 0.6 m) translate to molded parts of less than 18 × 12 × 12 in. (457 × 305 × 305 mm). In some cases, larger cores and cavities are created by producing multiple inserts and assembling them.

The greatest limitation of direct rapid tooling is surface finish. When applying rapid prototyping to direct tooling applications, secondary machining, grinding, and polishing

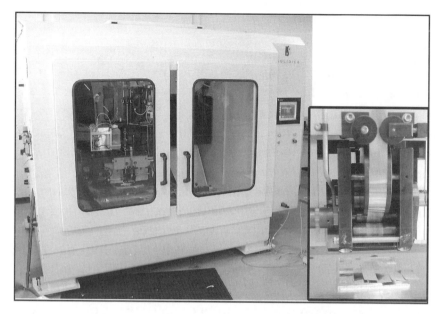

FIGURE 10-8. The Form-ation™ 2030 produces rapid tooling by ultra-sonically welding aluminum tape. *(Courtesy Solidica, Inc.)*

is commonly expected. Since the technologies are additive, stair stepping will occur in the tool (as it does with a prototype). Additionally, most of the processes used to fabricate metals lead to a rough surface that must be finished to allow molded parts to release from the tool and produce a cosmetically pleasing component. These secondary operations introduce time, cost, and quality variables into the tool creation process.

As described earlier, a rapid prototyping device's accuracy is most often a function of part size. As a part, or in this case a tool, enlarges its tolerance deviation increases. Since tools and tooling inserts are much larger and contain more material, the tolerance deviation of a tool will often be greater than the 0.005–0.030 in. (0.13–0.76 mm) of a prototype.

Future Developments

After digesting the limitations of rapid tooling, it would be easy to discount the application. However, this could be shortsighted, since research and development are under way that could change rapid tooling. Rather than being an alternative to machined tooling, the future advances may create a disruptive technology that changes the way tooling performs.

As an additive process, rapid tooling is insensitive to design complexity. This allows the creation of tooling with features that are impossible in subtractive processes.

With some innovation and research, rapid tooling holds promise as a solution that greatly reduces cycle time in the manufacturing process. When production volumes are high, the cycle time for manufacturing operations is a key determinant of cost and time. In terms of injection molding, cycle time is the rate at which molded parts are produced. In a production environment, cycle time is measured in seconds, not hours or minutes. With each second cut from the molding process, capacity increases, efficiency increases, and cost decreases to yield greater profit.

A reasonable reduction in cycle time can have a major impact on a company's bottom line. For example, if cycle times were reduced by 25%, and all other costs remained fixed, profit could improve by 100–200%. With this impact on operations and profitability, many research efforts have shifted their focus from faster tooling to faster cycle times. Rapid tooling holds promise in this area through two techniques: conformal cooling and gradient materials. Both could change the way production tooling is designed and constructed. When this occurs, rapid tooling will move from prototype development to full-scale production applications.

Conformal Cooling

Injection molds have cooling lines. These allow water to pass through the tool to remove heat. In doing so, cycle time is reduced. Yet, with conventional processes, these cooling lines are limited to straight pathways since they are machined into the tool. With straight channels, the tool cannot be cooled uniformly. To increase the efficiency of heat transfer from the tool, a new concept in cooling channel design is being developed. By using cooling channels that conform to the contours of the core or cavity, the rate of heat dissipation is increased and cycle times are decreased. This concept is called conformal cooling.

Conformal cooling takes advantage of the additive nature of the process to produce cooling lines that are nonlinear. Since the rapid tool is constructed one layer at a time, the conformal cooling path can be complex and convoluted. Some conformal cooling paths follow the core or cavity shape while others take on helical or spiral shapes. By introducing fluid flow consistently around the tool and within areas of high heat retention, the tool can maintain operational parameters at reduced cycle times. In *Figure 10-9*, a core with conformal cooling is shown next to a computer-aided design (CAD) illustration of the cooling channels within the core.

In the future, conformal cooling promises to be a powerful solution for production applications. The potential is so great that there is likely to be a new area of research and analysis that relates to the study of heat transfer characteristics and the impact on molding efficiency. Someday mold-flow analysis software may include the evaluation of alternative cooling channel design.

FIGURE 10-9. As shown in the CAD illustration (left), a core (right) with conformal cooling can improve heat dissipation to reduce cycle time. This core was produced by D-M-E MoldFusion™ with the ProMetal™ system. *(Courtesy D-M-E and Extrude Hone)*

Gradient Materials

Like conformal cooling, the use of gradient materials promises to reduce cycle times. However, it also has the potential to improve other tooling characteristics such as tool life.

Building from the additive nature of rapid tooling, gradient material tooling is constructed of multiple materials. However, there is not a discrete boundary between materials. Instead, one material gradually blends with another. Applying the materials in gradients eliminates the problems associated with dissimilar thermal expansion among different materials. When constructing the rapid tool, varying ratios of metal alloys would be deposited. For example, at the molding surface of the cavity, tool steel may be deposited. On the external surface of the tool, bronze could be deposited. Between these two extremes, the blended material would transition from a high ratio of steel to a high ratio of bronze.

The graduated placement of multiple materials can simultaneously address strength, surface wear, weight, and heat dissipation requirements. Tools constructed with gradient materials would increase efficiency and tool life while driving down cycle time.

RAPID MANUFACTURING

Today, few consider rapid prototyping as a viable option for manufacturing end-use products. Others view it as a possibility well into the future. However, necessity and innovation have already yielded beneficial rapid manufacturing applications.

Rapid manufacturing is a tool-less process. Direct from a digital product definition, production parts are manufactured without machining, molding, bending, or forming. Using the additive process can decrease time and cost while creating innovative processes and part designs.

There are few industries or applications required to meet specifications as stringent as those applied to military aircrafts and space vehicles. Thus, some find it surprising, even amazing, that rapid manufacturing has already been used to produce parts for fighter aircrafts, the space shuttle, and the space station. Fully qualified for flight, the rapid manufactured parts have yielded time and cost savings. However, for the limited number of units in production, tooling and molding were much more expensive and time consuming.

Coming down to earth, rapid manufacturing has been applied to other products with extremely low production volumes, such as race cars. Directly and indirectly, rapid manufacturing is used to construct metal and plastic components for NASCAR and Formula 1 race cars. In this fast paced environment where every ounce of weight reduction is critical, race teams have found that rapid manufacturing allows them to quickly realize production parts that improve performance.

Obviously, these representative examples are unique. Each has production runs measured in tens, not tens of thousands. And, each faces design challenges that are not common in the typical consumer or industrial product. Yet, everyday applications can benefit from rapid manufacturing. Innovative applications are arising every day as companies consider the advantages and possibilities rather than the obstacles and risks.

As more companies explore the opportunities, and as the technology develops into a suitable manufacturing process, rapid manufacturing will grow beyond a niche application to become a routinely used solution. New applications are likely to encompass two areas: customized products and small lot manufacturing.

Technologies

At present, there are no systems designed specifically for rapid manufacturing applications. Instead, many technologies that produce parts in a material suited for consumer use claim to be rapid manufacturing systems. All four technologies discussed in this book have evidence that illustrates their applicability to rapid manufacturing.

Stereolithography is used to mold custom, invisible orthodontic appliances and in the production of race cars. Selective laser sintering has been applied to aircrafts, space vehicles, and hearing aids. Fused deposition modeling has been used to replace assembly line components and battle-ready camera mounts for tanks. Using the same powder-binder printing technology that drives the Z Corporation process, companies have built their businesses around the rapid manufacture of ceramic filters and custom-dosage drug capsules.

The concerns, needs, and requirements of prototyping devices vary greatly from those used in manufacturing. In many cases, users have taken it upon themselves to modify the equipment or process to address manufacturing demands. What needs to happen, and will, is that new technologies and devices be designed and built specifically to serve the needs of the manufacturing community. As the number of applications grows, it is likely that such devices will be available for commercial use.

Limitations

There are currently many limitations to the application of rapid manufacturing. As with rapid tooling, the deficiencies have limited the use of the technology for production. Of the few successful applications, most are the result of the efforts of innovators and risk takers. Pragmatically, there is little reason to consider rapid manufacturing today. However, while maintaining a pragmatic view, there will be powerful benefits from the use of rapid manufacturing in the years to come.

To succeed, the limitations imposed by the current state of the technology must be overcome. The most obvious of these limitations is material properties. For the most part, rapid prototyping materials lack the mechanical, electrical, thermal, and chemical properties demanded of production devices. Many of the past rapid manufacturing successes have been built on systems that offer functional materials like acrylonitrile-butadiene-styrene (ABS) (fused deposition modeling) and polyamide-nylon (selective laser sintering). While these materials are functional, they do not deliver the properties of commonly used materials. Additionally, rapid prototyping as an industry lacks the breadth of materials and material properties required for broad-based use across many industries and many products.

Rapid manufacturing is gaining a tremendous amount of latitude in material development. Without the same constraints of traditional processes and with a diversity of processing methods, unique materials can be developed. It is possible that new classes of materials will develop with properties unmatched by traditional materials. Currently the industry is caught in a chicken and egg scenario. With the high cost of material development, it is difficult to justify work in this area when there are so few potential consumers. At the same time, many potential users exist, but the current state of materials retards the application of rapid manufacturing.

Another key barrier is output quality in terms of dimensional accuracy and surface finish. In many manufacturing applications, tight dimensional accuracy and cosmetically pleasing surfaces are demanded. As has been frequently mentioned, these are not traits of rapid prototyping systems. Unlike prototyping applications, secondary operations, to the extent that would be required, are not appropriate. Additional machining, sanding, and polishing of larger-volume production parts is not feasible due to time and cost. Past rapid manufacturing successes have been those that do not demand high tolerance or smooth-finish parts, or those in the one-off custom market where additional time and labor can be absorbed.

Although there are other barriers to rapid manufacturing, speed is the last one to mention. While RP is extremely fast for a handful of parts, when compared to machining or prototype molding, it is not fast enough for large-volume production runs. Many injection-molded parts are produced at rates measured in parts per minute, not per hour or day. While rapid prototyping eliminates the time for mold building, companies producing tens of thousands of parts a day can quickly reach and exceed a total throughput rate greater than that for rapid prototyping. While it is unlikely rapid manufacturing will be applied to razor blades and bottle caps in our lifetimes, it is probable that the speed will increase to a level suitable for low-volume manufacturing.

Benefits

When the limitations of the technology are overcome, and they will be, rapid manufacturing will offer tremendous benefits and create unthinkable applications. These gains will be achieved in two segments: customized products and low-volume manufacturing.

Mass customization was a term coined in the late 1990s. Although the term was intended to acknowledge the power of digital tools and the Internet, it was rapidly adopted as the ultimate application by the rapid prototyping community. Some consider mass customization to mean that every product, be it a telephone or a toaster, could be customized to the consumer's preferences while being produced economically and delivered quickly. While this may come true, the more likely scenario is that rapid manufacturing will be applied to those products that must be customized for user comfort and use. In this category, examples include hearing aids, artificial limbs, and orthotics.

The other application where rapid manufacturing will excel is in low-volume manufacturing. For the numerous products with annual sales volumes of 5,000–50,000 units, tooling is a major barrier. It is expensive when amortized over so few parts, and it takes a long time to receive. For many products, this eliminates new product launches, minimizes the number of product revisions, and reduces the ability to keep a product fresh in the consumer's mind. When rapid manufacturing succeeds, these barriers are eliminated. Companies with low-volume production demands will increase profitability while offering a continuous stream of new products.

Ultimately, high-volume manufacturers may use rapid manufacturing to change their business model. High-volume production runs, while extremely cost efficient on a per-part basis, carry high risk and high costs. Take for example inventory. A substantial percentage of a company's costs may be tied up in stockpiles of work in progress and finished goods inventory. This is one reason why just-in-time manufacturing was devised. Additionally, with large inventories, it takes a scientific process and intuitive effort to schedule and plan production so that demand is met while inventory is kept to a minimum. If a product does not sell well, a company may have to write off huge losses due to unusable inventory. With the tool-less approach of rapid manufacturing, it is feasible that high-volume production could shift to low-volume production of small lots. This would shift the expense of inventory and minimize the risk of carrying it. The lean inventory, rapid-response business process would give manufacturers unprecedented flexibility in production schedules and product revisions. Without the commitment to expensive tooling and large inventories, manufacturers could adapt to, and perhaps capitalize on, the ever-changing demands of their customers.

The most obvious benefit of rapid manufacturing is that the elimination of production tooling can reduce time to market by months. For high-volume tooling, it is common for delivery time to be two, four, or six months. In industries like computers, where time to market can be as little as six months to one year, the elimination of tooling would offer a 33–50% reduction in the cycle.

The final benefit of rapid manufacturing is that it unshackles the manufacturing floor from the constraints imposed by tooling. As rapid manufacturing further penetrates industry, creative and innovative approaches to production processes and product designs will surface. It is possible to envision a system that can produce a complete, multi-component product in one operation. Just think what the world would be like if "no assembly required" was a common practice on the production floor.

CHAPTER 11

Moving Forward

Rapid prototyping requires consideration. It is an amazing, powerful, and revolutionary technology for design and manufacturing. Lead times can be slashed, costs can be reduced, and quality can be improved. With these benefits, all organizations that design or manufacture mechanical components should consider rapid prototyping. However, consideration of the technology is not the same as adoption. As previously stated, rapid prototyping is not the best solution for all companies and every application, at least not yet.

While the industry has grown, it is not without challenges. The general consensus is that less than 20% of the design and product development community uses rapid prototyping. In the manufacturing and manufacturing engineering disciplines, the level of use is far less.

If the technology is so powerful, why do so few companies use it? Why were early predictions—phenomenal, rapid growth and the replacement of conventional processes—never realized? There are two components to the answer to this question. The first is a lack of awareness and understanding of rapid prototyping. The second is the combination of limitations and risks associated with the technology. Having read this book, the first obstacle is eliminated. Through the review and discussion of the technology, process, applications, benefits, and limitations, there is now awareness and understanding. And with this knowledge, informed decisions are now possible.

For many, the second obstacle, limitations and risks, may be insurmountable until the technology further develops and matures. To determine if this is a real barrier, an evaluation of the benefits, limitations, and risks associated with rapid prototyping should combine an understanding of the technology with the specific and unique circumstances within each company and for each project.

Having gained the knowledge and applied it to the needs of the business, many can make a final decision on the adoption of rapid prototyping. However, for the majority, the decision is likely to be that rapid prototyping may be a good solution, but it requires further investigation. This could be the best assessment, because a final, informed decision may require additional research and evaluation.

With the risks, efforts, and limitations that have been discussed, it becomes apparent that rapid prototyping is not a miraculous solution for the product development process. Instead, it is simply one option among the many tools available. For many, the limitations of the technology, in light of the expense and effort involved to achieve gains, is reason enough to forgo rapid prototyping. For others, the tremendous gains outweigh the effort and financial investment.

Whatever the decision, one additional factor should be considered—rapid prototyping is a relatively new industry. As a new industry, there are many opportunities for improvement, development, and advancement. These will decrease the limitations while growing the applications.

Consider rapid prototyping in contrast to computers. The first computer, the ENIAC, was announced in 1946. Developed at the University of Pennsylvania, the system was 100-ft (30-m) long, 10-ft (3-m) high, and weighed 30 tons (27,216 kg). It took another 19 years for the first personal computer to be developed. In 1975, MITS announced the Altair 8800. Fifteen years later, it was common to have PCs in the workplace, but they were limited in numbers. It was not until the mid-to-late 1990s that a PC on every desk was common in the business environment. The progress in computer technology, and the growth in its use, took nearly six decades.

Relative to the computer, rapid prototyping has developed at a much faster pace. However, like the computer, it will take years for the technology to mature and the rate of use to reach significant levels. There are real limitations and justifiable reasons to not use the technology. However, these will be addressed over the coming years. Through the next decade, rapid prototyping will change. There will be exciting developments in the technology, process, materials, and its application, ease of use, and price. What few realize is that many of these developments are already under way in research labs across the world.

With the coming change and with the growth in use that will accompany it, staying informed is critical. For those who choose to implement rapid prototyping and those who do not, the message is the same: keep abreast of the technology and its developments. Change is coming.

BARRIERS TO GROWTH

While many obstacles to the use of rapid prototyping are unique for each company, several are common for all companies and all industries. These obvious barriers to the growth of the industry also happen to be those receiving the most attention. As research and development address each of these obstacles, the likelihood of industry growth increases.

Computer-aided Design (CAD)

As described in Chapter 3, rapid prototyping demands 3D CAD input. Without the unambiguous, digital depiction of a three-dimensional solid or surface model, rapid prototyping is incapable of producing a part. A limiting factor to rapid prototyping's growth is that many do not use 3D CAD. Industry experts believe that 3D solid modeling users are in the minority. The vast majority of CAD operations continue to be 2D. Until these users convert to 3D modeling, it is unlikely they will incorporate rapid prototyping in the design process. However, software prices and price/performance of the computers to drive the applications have declined significantly in the past decade. This makes 3D CAD increasingly accessible to those in design and manufacturing, which should fuel the growth of 3D CAD use, and in turn, growth of the rapid prototyping industry.

Better Materials

Material properties, although greatly improved since the first "brittle" stereolithography resin, continue to be a limitation of rapid prototyping. When products are manufactured in thermoplastics like acrylonitrile-butadiene-styrene (ABS), polycarbonate, nylon, or polypropylene, there are no rapid prototyping materials that can deliver an exact match of all the mechanical, thermal, and electrical properties of the production resin. The same is also true for metals. While advances in technology now provide for rapid prototyping in steel, aluminum, and titanium, the variance in the processing parameters yields material properties that differ from the production materials.

To fuel rapid prototyping's growth, two things must happen. First, new developments must yield a broader range of materials and material properties that increasingly approach those of production materials. Second, the rapid prototyping industry must find new applications for the technology. For example, in some of the direct metal processes, the resulting properties of the material in the rapid prototype do not match the production materials— they exceed them. As materials develop and the breadth of selection grows, and as users find innovative applications for these materials, the adoption rate of rapid prototyping will increase.

Improved Output

While rapid prototyping offers speed and throughput, it is meaningless if the quality parameters of the prototypes do not match the application's requirements. Until such time that rapid prototyping can reliably and repeatedly produce prototypes, models, and patterns to the quality level commonly expected of conventional processes, there will be resistance to its use. However, the demands on the prototype must be reasonable and realistic. Many prototypes do not need to be held to tight tolerance or produced with a mirror-like finish. Inappropriate quality standards will undoubtedly reduce the number of rapid prototyping applications.

Measured against the quality parameters of early rapid prototypes, tremendous strides have been made. To varying degrees, the technologies have improved in areas such as dimensional accuracy and surface finish. Each year, new improvements are released. This trend will continue in the form of developments in current and new technologies.

Ease of Use

Uncrate it, plug it in the wall, and hit "start." This is the desire of the design and manufacturing community and a goal of each rapid prototyping manufacturer. Currently, many systems demand highly trained, highly skilled operators, and in some cases, extensive facility modifications.

There has been significant progress since the birth of the rapid prototyping industry. Systems are easier to operate and have fewer operational demands. But there is plenty of room for improvement. More companies will consider rapid prototyping when it becomes a push-button, plug-and-play device.

Lower Cost

A single design flaw that goes undetected could cost a company tens of thousands, even hundreds of thousands, of dollars. A decrease in the product development cycle of a single week can generate a like amount of additional profit. Yet, for many, it continues to be difficult to justify the expense of a rapid prototyping device and its ongoing operation. As with the computer, demand will grow as price decreases.

Prices have declined, especially in the 3D printer market. With some devices in the $30,000–70,000 price range, they are increasingly affordable and justifiable. However, prices will need to further decline for 3D printers and enterprise prototyping centers to make them affordable for the majority of designers and manufacturers. Additionally, the cost of materials, supplies, repairs, and general operation must also decline to fuel the rapid growth of the industry.

Cost, ease of use, quality, and materials are areas of improvement being addressed. However, the speed of advancement may be dependent on the rate of industry growth. Conversely, the rate of growth may be dependent on advances in these areas. It is expensive to design and develop rapid prototyping systems and materials, and the expense must be justified with sales. In light of this catch-22, the rate of technology advancement and industry growth may be relatively slow and incremental until dramatically different technologies, innovations, or applications emerge.

With the limited size of the rapid prototyping industry, some developments may not be financially justifiable. This is especially true for niche technologies and materials applicable to only a tiny segment of the market. For this reason, it is most likely that the bulk of developments will occur in general-purpose applications or those with substantial market potential.

FUTURE DEVELOPMENTS

The rapid prototyping industry is rapidly changing. Much has been accomplished since the 1987 introduction of stereolithography. However, this will pale in comparison to the developments that will arise in the short and long term.

There are multiple areas of research. Some organizations are focusing on overcoming the limitations of current technology. Others will focus on new applications and new technologies. Still others have set their sights on advanced applications like rapid tooling and rapid manufacturing. With the level of change that will come, those who use the technology and those who have elected not to will find that these developments will impact their futures.

It is imperative to continue researching and investigating the technologies of rapid prototyping. Equipment manufacturers, material suppliers, universities, research organizations, and end-user companies will contribute to advancements in rapid prototyping, rapid tooling, and rapid manufacturing. Over the next 5–10 years, this research will yield advancements that will make the application of these technologies more commonplace. As companies adopt these processes, they will fuel further research and development, and this will lead to even greater application of rapid prototyping, as well as rapid tooling and rapid manufacturing. In effect, once these rapid technologies reach a critical mass, the number of users and applications will explode.

Although it is impossible to predict the future with a high degree of accuracy, there are some forecasts of future developments that many in the rapid prototyping industry believe will become realities.

Standardization

The rapid prototyping industry lacks standardization. There are many methods of additively producing a prototype and many processes, and the output quality ranges widely. Without standardization, industry growth may be restricted.

Another important reason for standardization is that it brings a higher degree of confidence that obsolescence will not be an outcome. In an environment without standards, each technology is at risk of being superseded by a newer, better one. This risk of obsolescence causes many to delay a buying decision. Choice and selection are coveted. However, with a wide range of systems, outputs, processes, and materials to choose from, the evaluation and selection process can be daunting. Often, when faced with too many choices, the easiest decision is to do nothing.

Over time, the rapid prototyping industry will begin to standardize. Software architecture, operating methods, technology, and reasonable output will all become standardized. However, the standards will be decided by the purchasing decisions of the industry, not by companies that manufacture systems or develop software. As time passes, the leading technologies and processes, as measured by market share, will become dominant in the industry. In this capacity, each will contribute to the creation of standards. With this development, there will be greater ease and comfort in selecting a technology, which will fuel further growth.

It is likely that standards will not be industry wide. Instead, they will develop along the lines of applications and functionality within each class of rapid prototyping technology.

In Chapter 4, four classes of rapid prototyping technology were presented: 3D printers, enterprise prototyping systems, direct digital tooling, and direct digital manufacturing. While the industry agrees that there needs to be a distinction between low-end and high-end systems, there is no agreement as to what to call each. The class names proposed in this book are not intended to be the final word. Rather, they were proposed for clarity and further discussion.

As the technology further develops and matures, new classes and new labels will arise. Perhaps, as some have suggested, these classes may be called office and lab devices, 3D printers and productivity systems, or general-purpose and engineering resource systems. Classes may also develop along the lines of application. In the future, there may be biomedical systems, architectural systems, design systems, tooling systems, and manufacturing systems. As with computers, where there are servers, desktop, laptop, notebook, and handheld devices, the classes of rapid prototyping systems also will be segmented into sub-classifications. Along the lines of size, functionality, or application, the sub-classes will arise to give greater clarity of definition.

The problem in classifying systems today is that most are general-purpose devices. In the future, there will continue to be general-purpose machines, but more importantly, there will be devices designed for specific applications.

Specialization

When a tool is designed to do many things, it often performs each task to only an acceptable level. However, when a tool is designed to do one thing, it must do that one thing very well. In the future, the rapid prototyping industry will see the development of specialized devices. This will accompany the birth of companies that are equally focused on a specific application or industry. As illustrated in Chapter 4, technologies tend to have a specific area of excellence and application. However, they become general-purpose devices when the equipment vendors and users attempt to expand their breadth of application in the pursuit of market share and profit.

The rapid prototyping industry has yet to fully mature. Still developing, it is difficult to identify applications that hold great potential for market acceptance. In light of this forecasting challenge, it appears that many of today's rapid prototyping technologies are attempting to improve the odds of success by promoting the technologies for a broad range of applications. In effect, they are not investing in the development of an application-specific technology until it is determined that the application holds enough potential to make it worthwhile.

Outside the product development community, additive processes are already being researched and developed to address specific applications. Some have elected to focus on biomedical applications and others on small-scale (nano) part production. Others have elected to focus on tooling and some on jewelry. Within the product development community, some are beginning to realize that a jack-of-all-trades is a master of none. These companies are the pioneers of rapid prototyping specialization. And it is likely that the specialized equipment serves the user better.

As indicated in Chapter 9, when evaluating rapid prototyping it is best to narrow the application of the technology to those areas where it is most needed and where the most benefit is derived. This is an alternative way of saying that the technology selection should be focused on specific needs and desired benefits. If trying to satisfy all applications with a single technology, it is likely that most applications gain marginal value and few applications realize the full potential of rapid prototyping.

As the future unfolds, technology will become specialized for the intended application. With this focus, even greater advances will result, since the needs of the application, not the needs of an entire industry, are identified and addressed. As discussed in Chapter 10, rapid manufacturing is a clear example of this concept. Using general-purpose devices in a manufacturing process is ill advised. For rapid manufacturing, new equipment designed from the ground up must be targeted at rapid manufacturing to address the needs of the production environment.

3D Printing and Rapid Manufacturing

Two areas of promising growth are 3D printing and rapid manufacturing. As the future unfolds, the industry will realize powerful developments in both areas, and as a result there will be significant growth.

When considering the full scope of rapid prototyping—as the entire collection of additive technologies—3D printing and rapid manufacturing applications are likely to receive the most attention. While a given product may have a few prototypes, it is likely that there are many more concept models and thousands more manufactured items. The sheer volume of parts for these two applications is likely to invite the focus and specialization of many technologies and companies. This focus will then yield significant advancements.

3D Printing

While some systems loosely fit the description of a 3D printer, future advances will make these devices what the consumer wants them to be: low cost, easy to operate, and fast. Additionally, the technology of the future promises to provide a true office-environment device.

In the short term, it is unlikely that RP will advance to the point of being a low-cost, desktop device that becomes the 3D version of an inkjet or laser printer. Likely, the technology will develop so it is appropriate for an office environment where it is a shared resource of the engineering and manufacturing departments. In this role, the 3D printer will be analogous to the departmental copy machine. To this end, several obstacles will be overcome, including:

- The purchase price and operating expense must decrease.
- Systems must decrease in physical size to that of an office machine.
- To be practical as a shared device, the systems must be easily installed and operated without vendor training and support or a dedicated operator.
- Noise levels of the devices must decrease to the point where they are the same as that produced by a copy machine.
- All systems produce some dust or debris that makes them better suited for a lab or shop environment. This must be remedied.

These advances will be made. They will open a big market for rapid prototyping systems that produce conceptualization tools ideally suited for the office environment. What would an office or department be without a copier? Inconceivable, isn't it? At some time in the future, the same will be true for the 3D printer.

Rapid Manufacturing

Very few companies have applied rapid manufacturing to the production of finished goods. For some, rapid manufacturing is envisioned in the future; others remain convinced that the dream will never be realized. While it may be five, ten or twenty years for the vision to be realized on any sizable scale, rapid manufacturing will become a reality at some point in the future. The advantages of a tool-less operation are too big to ignore.

Obviously, rapid manufacturing eliminates the time and cost to produce tooling, and that has tremendous benefits. But rapid manufacturing's benefits go much farther than that. With rapid manufacturing, many aspects will change.

With an ability to make an economic order quantity equal to one unit, the possibilities are endless. The elimination of tooling could usher in an age of product customization not currently feasible.

The consumer and business marketplaces could see a significant rise in new products since the obstacle of tooling investment for short-run products is eliminated. There are only 500 Fortune 500 companies, but there are hundreds of thousands of small companies with great ideas that can be unleashed with the elimination of tooling.

Rapid manufacturing will enable the elimination or minimization of the greatest drain on any organization, inventory. When products can be made on-demand, in small lots, there is greater latitude for new product development. Selling current inventory is not an issue. There are huge reductions in the expense of carrying inventory and a corresponding decrease in facility size since the inventory stocking levels will be minimal.

There are many more benefits of rapid manufacturing, but before any are realized on a large scale, barriers must be eliminated. Currently, the devices are prototype tools that lack the controls of viable production devices. Likewise, most of the available materials are best suited for product development applications. While some rapid prototyping technologies deliver properties that approach or mimic those of an injection-molded part, most do not. To lead the way to rapid manufacturing, systems will be redesigned as production devices, and new classes of materials, including plastics, composites, and metals, will be developed.

Rapid manufacturing has the potential to change the way manufacturing is done. While change is the only constant, and change is for the better, the scope of change to implement rapid manufacturing will create another obstacle. Companies, departments, and individuals must be receptive to this change for it to be realized. Many will not be open to the change for fear of risk and impact on the company and individual roles. There will be no easy solution to opening up the market to these changes. Likely it will take time, much more than was required for the technology to develop or the concept to be accepted.

At some time, rapid manufacturing will be a commonly applied technology to custom-made and low-volume products. But to achieve this there will need to be changes in technology, materials, processes, perceptions, and goals.

CONCLUSION

Rapid prototyping is a relied upon tool in thousands of organizations around the world. From aerospace to consumer electronics, this collection of additive technologies has demonstrated a unique ability to rapidly construct models, prototypes, patterns, tools, and finished goods without regard to design complexity. Finding errors and flaws, detecting areas for improvement, and discovering cost-saving measures are just a few of the advantages of rapid prototyping that companies have come to rely on.

As improvements are made, applications discovered, and new technology announced, the barriers to the use of RP will diminish. The technology will become ubiquitous. It will become as commonplace as CAD and computer-aided manufacturing (CAM), and quite possibly as commonplace as the copy machine or printer. When this happens, more companies will directly benefit from RP. Until that time, realize that the technology has touched each of our lives. The cars we drive, the computers we use, and the home appliances we count on have been made better, brought to market faster, and produced cheaper with rapid prototyping. Industry has also benefited from rapid prototyping through advances in other competitive technologies. This has been fueled by the perceived competitive threat in the areas of time reduction and capability improvements. For many engineering and manufacturing processes, what once took weeks or months now can be completed in days or weeks.

"There are risks and costs to a program of action. But they are far less than the long-range risks and costs of comfortable inaction."
John F. Kennedy

Rapid prototyping will continue to grow and flourish. The technology is here to stay. The proven concept of construction through additive, layer-based processes will continue to spawn new ideas and developments in other industries and applications. Printing of human tissues, construction of buildings, and plotting of landscapes are just a few of the related applications that rapid prototyping for mechanical components has launched.

New applications will require change. Change within the technology, materials, industry, and even within ourselves. While rapid prototyping will change, it is also a tool for change. Rapid prototyping finds application in a business environment of decreased resources and cost control that at the same time demands decreased product development cycles and improved quality. Unleashing innovations in design and process, rapid prototyping can foster other changes within the product development and manufacturing disciplines. However, this will require an openness to change on the part of each individual. The risks and efforts of implementing rapid prototyping, and the change that comes with it, must be welcomed.

In consideration of all companies and all products, the soundest advice offered is to continue to question, evaluate, and consider rapid prototyping. Combine this investigation with an open mind to the possibilities of the technology. This is true for those who choose to implement the technology and those who do not.

For those who are beginning the process of justifying and evaluating rapid prototyping technologies, there are many techniques and processes to review. While the discussion in this book has centered on stereolithography, selective laser sintering, fused deposition modeling, and powder-binder printing, be aware that there are nearly two-dozen technologies available. The exclusion of all other processes was done for clarity and brevity. Yes, each of the four processes discussed are leading technologies. Yes, each has the potential to satisfy the needs of an organization. However, the other rapid prototyping technologies are as likely to satisfy individual needs and goals. A sound evaluation and a strong operation will be based on consideration of the entire range of processes. To assist in the quest for further information, Appendix C offers a listing of resources that may be beneficial.

Whatever it is called, rapid prototyping, freeform fabrication, or additive manufacturing, the process and benefits are clear. Rapidly producing a physical, 3D model without sensitivity to design complexity fosters improved communication, productivity, and efficiency. In an age of better, faster, and cheaper, rapid prototyping is a tool that enables individuals, departments, and companies to achieve these goals.

Remember, rapid prototyping is still just a tool. It complements the other tools available for design and manufacturing. Yes, it is a powerful tool, but it is only one of many.

BIBLIOGRAPHY

Institute of Electrical and Electronics Engineers (IEEE) Computer Society. uww.computer.org. Washington, DC: IEEE Computer Society.

APPENDIX A

Case Studies

RAPID PROTOTYPING PROMOTES MORE DESIGN ITERATIONS, SMOOTHER PILOT TEST

Technology: fused deposition modeling

Applications: form, fit, and function

Contributed by Stratasys, Inc. (Eden Prairie, Minnesota)

Oreck Corporation (New Orleans, Louisiana), a manufacturer of commercial vacuums, built a functioning vacuum assembled from rapid prototypes. Each of the 31 injection-molded parts was built with fused deposition modeling (FDM) in acrylonitrile-butadiene-styrene (ABS). With the technology, the company trimmed five months from the product development process and netted several hundred thousand dollars in savings on production tooling.

Since 1993, Oreck had used service bureaus for its rapid prototyping needs. Due to cost and time, the company would construct only a limited number of prototypes. At that time, it would take 7–10 days to receive a prototype order. To decrease time and cost, the company purchased an FDM Quantum™ in 1998.

Oreck used the fused deposition modeling system during the development of its new Dual Stack® commercial vacuum. Applying rapid prototyping to form, fit, and function testing, the company averaged six to eight iterations of each of the 31 injection-molded parts, as illustrated in the exploded view in *Figure A-1*. According to Rich Conover, Director, "Without using the FDM RP, several design itera-

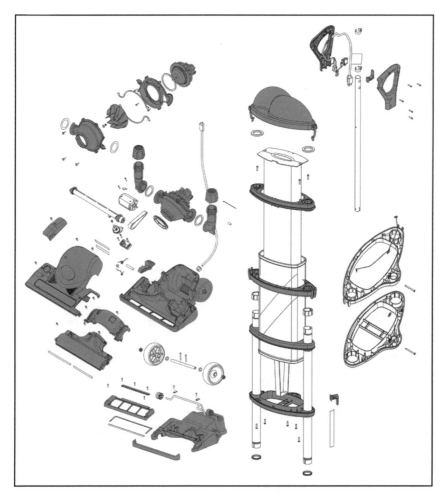

FIGURE A-1. Exploded view of the Dual Stack vacuum. *(Courtesy Stratasys, Inc.)*

tions would have been very difficult and would have extended the development time considerably." Conover estimates that 95% of the fit problems were caught prior to tooling.

When the design engineers were satisfied that they had made the necessary improvements to each component, the company designed and manufactured the injection-mold tooling, and an engineering pilot was done. Conover commented, "For the first time since I've been in this industry, all of the parts fit together during the pilot, which is not typical of injection-mold tooling." Assembled units from the engineering pilot undergoing testing are shown in *Figure A-2*.

The Dual Stack went into full production in the spring of 2000. By using rapid prototyping, Oreck reduced the time to market by four to six months. By catching problems early in the design cycle, before machining production tooling, the company estimates it saved several hundred thousand dollars.

FIGURE A-2. Using rapid prototyping to detect design flaws early, there were no fit problems in the engineering pilot of the Dual Stack vacuum. *(Courtesy Stratasys, Inc.)*

SNOWSHOE DESIGN HITS THE FAST TRACK

Technologies: selective laser sintering, stereolithography

Applications: form, fit, function, patterns, sales proposals

Contributed by Accelerated Technologies (Austin, Texas)

Spring Brook Manufacturing (Grand Junction, Colorado) planned to introduce an innovative recreational snowshoe design and offer it through the catalog of the world-famous outfitter, L.L. Bean.

Spring Brook needed to streamline its development process to accelerate market introduction. The snowshoe company needed the service of a company that offered rapid prototyping and room-temperature vulcanized (RTV) molding services to help it create functional, effective, prototypes of the snowshoe deck and binding.

The first snowshoe deck prototypes were produced using selective laser sintering technology with DuraForm™ GF, a glass-filled polyamide that produces high-definition features, a smooth surface finish, and enhanced mechanical integrity. Accelerated Technologies created two prototypes in just three days.

Next, RTV molds were created from the selective laser sintering prototypes. From the molds, cast urethane components were produced. To yield the look and feel of the production material, a rigid, translucent urethane was used to produce multiple deck prototypes. For the bindings, a master pattern was created in stereolithography to generate the RTV mold. The first urethane prototypes were cast in a Shore A 80 material that mimicked the soft-touch, low durometer production bindings. After evaluating these prototypes, a second set of bindings was cast in Shore A 90 material to increase the binding's stiffness. The entire process of creating patterns, molds, and cast urethane parts was completed in just two weeks.

The prototypes were used for design review with L.L. Bean, which agreed to offer the new snowshoe in its fall catalog. L.L. Bean needed the product photos by May of that year—long before production bindings were available—so stereolithography prototype bindings were paired with the production decks to create the persuasive image vital to creating consumer interest. *Figure A-3* shows a production snowshoe.

The urethane castings were used for "on snow" functional testing. After the testing was completed, Spring Brook released the tooling order for the snowshoe decks. In just six weeks, the production tools were completed, and the first injection-molded decks were delivered. The selective laser sintering models were also used to display the product at the first industry trade show of the season.

FIGURE A-3. Production Sagauche snowshoe, which was prototyped with stereolithography and selective laser sintering. *(Courtesy Accelerated Technologies)*

FINITE ELEMENT ANALYSIS RESULTS PRINTED IN 3D

Technology: powder-binder printing

Applications: concept models, presentation aids

Contributed by Z Corporation (Burlington, Massachusetts)

Constar (Alsip, Illinois) creates a wide range of plastic packaging, which is primarily used for beverages. The company takes responsibility for a product at one of several phases in the development process, from earliest concept development to implementation of fully developed designs.

Prior to rapid prototyping, Constar would work with various forms of finite element analysis (FEA) output to help the customer understand the need for change. However, describing the issues in words supported with 2D color prints often left customers disappointed. They would hear that their design would require a change, without really understanding and appreciating the issues behind it. To combat this, Constar uses Z Corporation's color, 3D printing solutions.

In Constar's experience, using 3D color appearance models to represent FEA data to the various constituents in the design process is very effective in identifying problems, demonstrating the need for change, and getting consensus on improved designs. More effective communication helps strengthen customer relationships and ultimately makes better products.

Vacuum Effects on Hot-fill Container

Constar uses FEA to test the effects of vacuum for hot-fill containers. Hot-fill bottles are filled with a liquid at or above 176° F (80° C) to sterilize the contents. When a liquid or semi-viscous product is heated to this temperature, the volume expands. For many beverages, this expansion is just slightly over 3%. After the products are packaged, they are cooled with a water spray. The product volume then decreases, causing a vacuum to form and the bottle to deform inward. If the vacuum effects pull the bottle out of round, this negatively affects labeling, storage, shelving, and customer perception.

To combat the effects of vacuum, bottles are thermally stabilized and designed with vacuum panels. Designed correctly, the panels will expand and then contract without affecting the geometry of the rest of the bottle.

For one bottle design, Constar's FEA analysis revealed a problem. The vacuum panels were not the only areas that were deforming. Realizing the need for a design change, the company printed the part in color and used it to explain the problem to the customer and propose the solution. The design problem was immediately clear, and the customer agreed to the necessary design changes the same day.

Buckling Under Top Load

Constar also uses FEA to check for buckling, which is when the plastic creases as a result of a top load. Bottles are exposed to a top load when the container is filled and capped and then stacked in several layers of boxes. A poorly engineered bottle deforms under top loading, which permanently creases the plastic, and presents a flawed product to the customer.

In another project, the color 3D print clearly showed the impact of the top load. The model illustrated color surface data and the resulting shape of the bottle. With the physical model, the customer had a very tangible sense of the problem and was able to quickly agree on a good solution.

ANCIENT GREEK STATUE REPRODUCED WITH RAPID PROTOTYPING

Technology: stereolithography

Applications: art, presentation

Contributed by Materialise N.V. (Leuven, Belgium and Ann Arbor, Michigan)

A remarkable statue is standing in the hall of the Materialise headquarters in Leuven. It is a reproduction of the ancient Greek Kouros sculpture that stands nearly 6 ft (1.8 m) tall. The statue is the result of the EcoMarble project, which included the investigation of processes for the capture of digital data and reproduction of sculptures. Participants in the project included Materialise N.V., GEO-Analysis, Focke Museum in Bremen, Fitzwilliam Museum of Archaeology in Cambridge, and the Archeological Receipts Fund in Athens.

The aim for Materialise was to make a life-size replica of a Kouros sculpture. A Kouros is considered one of the most distinctive products of the Archaic era, the period of ancient Greek history from about 650–500 B.C. In this time period, life-size Kouros sculptures marked graves and stood near temples as gifts to the gods. A Kouros sculpture was always an athletic-looking male nude standing in a frontal pose. His left foot was slightly forward; his arms held close to his sides; and his hands were clenched. Kouros sculptures were a general representation of youth and strength, not portraits of people.

GEO-Analysis scanned a Kouros sculpture and delivered the data to Materialise. The preparation of the scanned data took one day. Materialise then built the Kouros sculpture on the company's proprietary Mammoth II® machine. This stereolithography machine, with a build area of 85 × 24 × 20 in. (216 × 61 × 51 cm), built the full-size statue with a layer thickness of 0.006 in. (0.15 mm). In just under 100 hours, the stereolithography reproduction of the Kouros sculpture was complete.

With a volume of 610 in.^{3} (10 L) and a wall thickness of 0.2 in. (5 mm), the Kouros sculpture used 35 lb (16 kg) of PolyPox® resin. The weight of the statue after removing the support structures was nearly 24 lb (11 kg). After covering the statue with a ultraviolet (UV) paint, Kouros was ready to be displayed. Standing steadily on its feet, the prototype statue—perfectly illustrating Materialise's giant stereolithography capacity—would become the center of curiosity for visitors to its headquarters and booths at fairs all over Europe.

TECHNOLOGY AND INNOVATION USED TO BOOST AGING PRODUCT

Technologies: fused deposition modeling, computer numerical control (CNC) machining

Applications: form, fit, function, communication aid, marketing materials

Contributed by Leyshon Miller Industries, Inc. (Cambridge, Ohio)

Fomo Products, Inc. (Norton, Ohio) is a manufacturer of polyurethane foam systems for the building and construction industry. The company's one and two-part kits are available as both disposable and refillable units. One of their products, Handi- Foam® 2-12, presented the company with a tough decision when the 10-year-old tooling neared the end of its useful life. While the product had been highly successful, annual sales had dropped to 50,000 units from its peak of 120,000 units. The options were clear: scrap the product, repair the tooling, or redesign the product and retool.

To help Fomo select the right path and devise a strategy, the company called on Leyshon Miller Industries (LMI). Fomo knew LMI had the talent and experience to assist them in the decision and execution of the design and manufacturing work. Jointly, the companies elected to redesign and retool Handi-Foam 2-12.

The project began with the redesign of the package. After submitting design concepts, Fomo's European division requested a radical styling change, one with a "stealth-like" appearance. With a dramatic departure from previous product styling, LMI knew it could not rely on CAD files and engineering drawings to communicate the design intent. To describe the new design, LMI elected to use photo-realistic rendering tools, as shown in *Figure A-4*. While rapid prototypes would have been the best choice to illustrate the design—time, distance, and cost were barriers.

The redesign needed to be reviewed and approved by Fomo's U.S. and European operations. To construct rapid prototypes would have taken 10 days. With the additional time for overseas shipment, the pace of the project did not accommodate the time required for rapid prototypes. To overcome these challenges, LMI imported the CAD data into Alias® to generate the photo-realistic images. While these did not allow the customer to hold a "real" part, the images satisfied the primary goal of styling review. Additionally, the project plan included physical prototypes after the styling was complete. Thus, the team was comfortable in the review of the renderings, since the physical prototypes would allow a full evaluation later in the project. Two renderings cost $1,500 and took three days to generate.

The next step in the project was the mechanical and functional design of Handi-Foam 2-12. As identified in the stra-

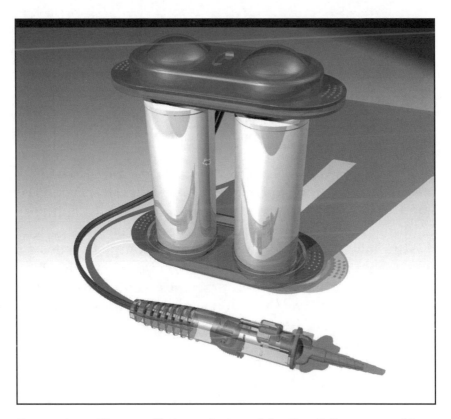

FIGURE A-4. Photo-realistic rendering of the Handi-Foam 2-12. *(Courtesy Leyshon Miller Industries)*

tegic planning meetings, for the consumer market, the product would have to be simple to use. Two product innovations emerged in the areas of the dispensing mechanism and the process for product activation. For the dispensing mechanism, a unique two-piece valve was designed. The other design innovation was the activation of the two-part system. In the old product, a knob was tightened to depress the valves on the canister. In the new design, the valves are engaged by pushing the housing down onto the canisters. Once activated, the valves are captured with snaps. Both of these concepts required form, fit, and functional analysis prior to production tooling.

As the styling design was completed and the mechanical design was in process, Fomo's marketing department requested samples for sell sheets and brochures to issue with the product launch. Having no production units, the team turned to rapid prototyping. Using a Stratasys fused deposition modeling system, all plastic components for the assembly were prototyped. These were finished and painted to give the appearance of a production unit, as shown in *Figure A-5*. For this application, LMI also considered CNC machining and stereolithography. However, FDM rapid prototyping was selected because of LMI's ability to build the prototype within a week at a lower cost.

With new, innovative features and a complete redesign, functional testing was a necessity. To construct the function prototypes, LMI used three processes: rapid prototyping, CNC machining, and injection molding. The selection of the process was based on its capabilities, time, and cost. Fused deposition modeling was used for the upper and lower housings, dispenser handle, and handle cover. CNC machining was used for the snap features, brass valves, and the prototype injection mold. The prototype injection mold was used for the valve's elastomeric stem. The combination of functional prototypes from these processes is shown in *Figure A-6*.

The successful launch of Handi-Foam 2-12 can be directly linked to the prototypes constructed to prove the product's form, fit, function, and appeal. This was most apparent when dealing with issues that could not be evaluated in a CAD environment. For example, the precision of the sealing features required hands-on experimentation. Using the physical prototypes to analyze the seals allowed multiple iterations to achieve the right fit. These modifications were then incorporated in the prototype tooling. LMI estimates these iterations saved Fomo $5,000–10,000 and weeks of modification time, while avoiding changes to the production tooling. Another example was the evaluation of

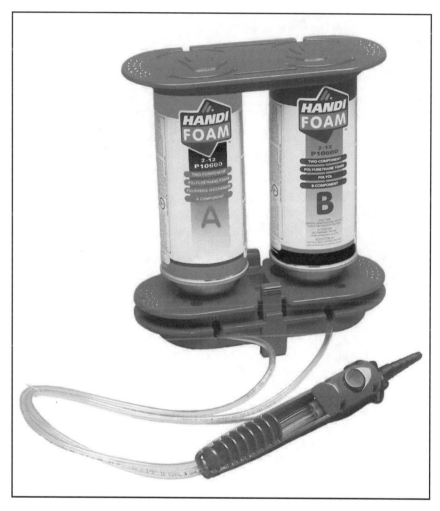

FIGURE A-5. When sales literature was needed, Fomo Products used fused deposition modeling for the product photography. *(Courtesy Leyshon Miller Industries)*

snap-fit feature stiffness. Determining what was too much and too little required hands-on testing. Holding the unit and manually engaging the snaps confirmed the functionality. CNC machining of the snaps minimized production tool iterations, which yielded estimated savings of $10,000–15,000.

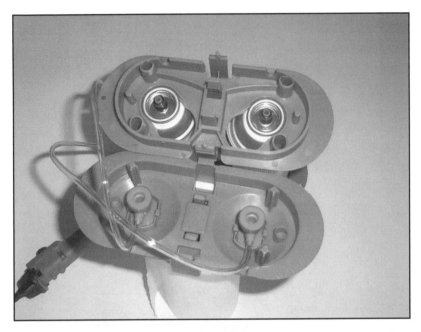

FIGURE A-6. A fully functional Handi-Foam 2-12 unit was produced with a combination of rapid prototyping, CNC machining, and injection molding. *(Courtesy Leyshon Miller Industries)*

The unit is now in full production and has met with very positive customer response. Fomo plans to implement many of the new features of Handi-Foam 2-12 in future designs of other products.

PROCESS COMPARISON: STEREOLITHOGRAPHY AND RTV MOLDING

Technologies: stereolithography, RTV molding

Applications: functional analysis and testing

Contributed by DSM Somos (New Castle, Delaware)

General Pattern (Blaine, Minnesota), a rapid product development service bureau, compared cost and time for the creation of an automotive register in stereolithography and RTV molding. The results of the analysis suggest that parts manufactured with stereolithography are often a more economical and faster solution for form, fit, and function reviews.

The register assembly was made of 13 components, including a housing, damper arm, thumbwheel and five vanes. In all, 10 assemblies were made with RTV molding and stereolithography.

Considering pattern development, mold creation, and part casting, the RTV molding process took 10 days to complete at a cost of $10,300. In contrast, the stereolithography process—two runs of an SLA® 5000 with DSM Somos Raven™ 7620 resin—took just three days and cost only $5,295. For 10 complete sets, the stereolithography process took 1/3 the time and cost half as much. *Figure A-7* shows an assembled register made by the stereolithography process.

Using General Pattern's data, the RTV molding solution offers a cost advantage for quantities in excess of 35 units and a time advantage when more than 80 sets are produced. However, the typical life of an RTV mold is 50 or fewer castings. With the requirement of new tools to produce larger quantities of castings, the time and cost advantages of stereolithography would prevail for much larger quantities.

Commenting on the results of the analysis, Denny Reiland, General Pattern's North American general manager, stated, "It is clear to us that designers, engineers, and original equipment manufacturers (OEMs) should be alert to the emergence of new materials that offer the potential of improving cost and turnaround time over traditional methods of modeling and prototyping."

FIGURE A-7. General Pattern found that this 13-component automotive register could be made faster and cheaper with stereolithography. *(Courtesy General Pattern Company)*

RAPID PROTOTYPING ASSISTS SURGEONS PLANNING SEPARATION OF CONJOINED TWINS

Technologies: powder-binder printing, stereolithography

Applications: medical modeling, pre-surgical tools

Contributed by Medical Modeling LLC (Golden, Colorado)

The use of rapid prototyping for anatomical structures is growing among the surgical community. Their use in such operations as total hip replacement and neonatal airway decompression has led to applications in some of today's most complicated surgeries. The benefit comes primarily to the surgeon, allowing true comprehension of the anatomy in 3D. This increased understanding leads to simulation of surgery on the model, including customization of devices. The quantifiable benefits of this pre-surgery planning include reduced surgical time and reduced anesthetic time. Given that operating room time costs an average of $100 per minute, time is money. Time is also life, as pediatric patients run an increased risk of complications with anesthesia during longer operations.

Surgical separation of conjoined twins has been described as "the most complicated surgical procedure known to man." Massive teams of surgeons, nurses, anesthesiologists, and caregivers are carefully coordinated during these surgeries that can last days. In 2001, separation of Nepalese twin girls, conjoined at the head, took over 100 hours in Singapore. Based on the sheer complexity of these cases, they require an extreme amount of pre-planning and the use of every tool available to allow for visualization of the anatomy. In the Singapore procedure, the team of surgeons, lead by neurosurgeon Dr. Keith Goh, used exotic surgical simulation virtual reality equipment developed specifically for the case. The team also used rapid prototyped anatomical models of the girl's skulls, including their brain vasculature, for identification of critical areas of conjunction.

It is hard to imagine how long the surgery would have taken without the physical models, or more importantly, if the surgery would have even taken place. For a typical surgical case involving a model, the decision has been made to proceed with the surgery long before the model is ordered. Increasingly, the decision to proceed is made after consultation of advanced medical imaging study results and rapid prototyping models.

Rapid prototyping models are typically produced using a computed tomography (CT) scan of the bone structure. The involvement of multiple types of tissues in cases of conjoined twin separation has led to use of different imaging technologies plus rapid prototype generation techniques. One critical area of concern for almost all twin separation cases is the amount of skin coverage. Where two objects are conjoined and separated, they leave an area that must be covered by skin. Without proper closure of the wound, the patient may develop an infection and die.

The Ibrahim Twins

Mohamed and Ahmed Ibrahim were born in Egypt in June of 2001. These twin boys are craniopagus twins, which means they are attached at the head. Soon after their birth, the boys were moved to the most advanced hospital in the country at the University of Cairo. Under the care of pediatric surgeon Dr. Nasser Abdel-Al, the infants were cared for until the age of six months. During their first six months of life, surgeons at the hospital searched the world for a team of surgeons who would evaluate the twins for possible separation surgery. Complicating this search was the fact that over the last 30 years only 30 sets of craniopagus twins had been separated, and only a handful had survived without death or major neurological complications.

Dr. Nasser found a surgeon willing to evaluate the twins in Dallas, Texas, Dr. Kenneth Salyer. Dr. Salyer is a world-renowned craniofacial surgeon who was previously involved in the successful separation of twin girls conjoined at the head. In June 2002, the Ibrahim twins arrived in Dallas. The World Craniofacial Foundation (Dallas, Texas) agreed to coordinate care and raise funds for the twins journey from Cairo.

Upon arrival in Dallas, the twins underwent a series of imaging procedures geared at giving the surgeons all the information possible about their conjunction. Imaging studies revealed that the twins skulls were completely fused together and their brain blood vessels (or vasculature) were heavily intertwined. MRI studies revealed that they had largely separate brains, connected at about 10% of the surface. Dr. Salyer contacted Medical Modeling prior to the twins arrival to enlist support for physical modeling of their condition. Imaging studies performed in Dallas were immediately sent to Colorado in preparation for physical modeling using rapid prototyping.

Of critical importance to the team evaluating the twins was the brain vasculature and its relation to surrounding bone structures; their brains and the anatomy of their conjunction; and the external soft tissue structure. To generate physical models of these structures, there must be an understanding of the nature of the imaging studies used as the basis for the files needed to make the rapid prototyping models.

The CT scanning, while great at distinguishing bone from surrounding tissue, is not suited to detailed 3D visualization of the brain or vasculature. Introduction of a contrast agent into the bloodstream before the scan allows for better visualization of the vasculature system. Therefore, the CT provided information about the bone, skin, and vasculature. MRI and functional (moving projection) MRI were performed for the specific task of visualizing the brain, vasculature, and real-time flow of blood through the brain.

In craniopagus twin separation surgeries, the critical factor often comes down to how much interdependence there is in the vascular system. In the case of the Ibrahim twins, the functional MRI study revealed that they shared much of the venous drainage structure, a dangerous condition.

Further complications arose from the fact that the twins were only six months old and were growing rapidly. This could have a very detrimental effect on surgical preparation where every millimeter counts.

Image processing at the modeling facility was the first step to making the rapid prototype model. Since MRI and CT data are 2D imaging modalities, specialized software was needed to read, process, and export useful data in 3D formats. Mimics™ (Materialise, Ann Arbor, MI) software was used to import and process the images of the two primary data sets. While some of the objects could be semi-automatically segmented, several of the objects required manual segmentation by a technician with a skilled eye. STL files were output using CTM™ (Materialise, Ann Arbor, MI) software and used for registration of the objects.

Production of the models was done using stereolithography and 3D printing. Models of the bone and vasculature were created using stereolithography in Stereocol™ (Vantico, East Lansing, Michigan) material. Stereocol is the only resin allowing for embedding of a second color within a clear model in stereolithography. The vasculature was highlighted in red within the bone, as shown in *Figure A-8*. The 3D printing process, as commercialized by Z Corporation (Burlington, Massachusetts), was used to produce solid models of the skin and brains. The brains were also disarticulated, allowing surgeons to visualize the shared anatomy. Brain models were produced on a Z Corporation system in a starch-cellulose powder that was infiltrated with a flexible urethane, making the brains flexible. The craniofacial surgeons, pediatric neurosurgeons, and pediatric neuroradiologist all utilized the physical models in the evaluation of the case.

The first step in most surgical separation cases involves stretching the skin to allow an adequate amount to close over the defect left by the separation. This is done by inserting small, deflated balloons under the skin. Over time these balloons are inflated with fluid, and the skin grows and stretches in this area. For the Ibrahim twins, this posed many challenges because the twins were unable to stand. This complicated the expansion of skin in the area on which they laid. To aid in solving this problem, Dr. Salyer turned to Medical Modeling to produce a life- size model of the twins' skin structures, as shown in *Figure A-9*. This model, produced using 3D printing, would be used by

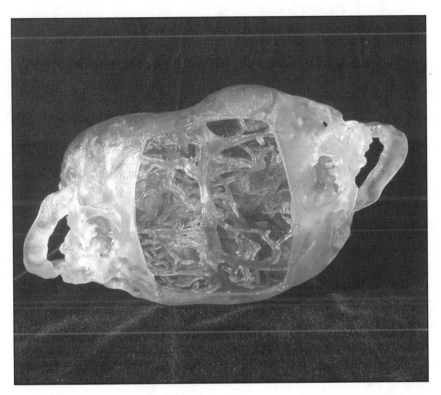

FIGURE A-8. Using stereolithography with Vantico's Stereocol™ resin, the model allowed surgeons to visualize the challenge by illustrating the shared bone and brain vasculature. *(Courtesy Medical Modeling, LLC)*

another company to create a custom jacket, shown in *Figure A-10*, which would provide padding in the shoulder areas, allowing the twins to lie down without interfering with the skin expansion. A model made with plaster powder, again with a Z Corporation system, was produced in four pieces and assembled for this purpose.

On April 28, 2003 the surgical team, led by Dr. Salyer, placed tissue expanders under the twins skin. On October 11, 2003 the twins were successfully separated. Aided by the rapid prototyping models, the 26-hour surgery went smoothly.

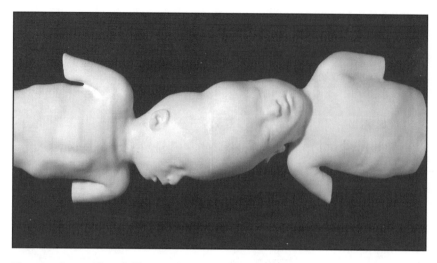

FIGURE A-9. This full-size skin model of the twins was produced on a Z Corporation 3D printer for the construction of a custom "jacket." *(Courtesy Medical Modeling, LLC)*

RAPID TOOLING CUTS PROCESSING TIME BY 50–75%

Technology: selective laser sintering

Application: tooling

Contributed by 3D Systems (Valencia, California)

Apex Mold and Die Corporation (Endeavor, Wisconsin) designs and builds plastic injection molds for a variety of OEM customers. The company specializes in producing high-tolerance molds for high-volume production runs. With their customers aggressively seeking to reduce time to market, Apex sought new solutions and found that selective laser sintering for rapid tooling was the answer.

Prior to its selective laser sintering (SLS®) system, Apex required four to eight weeks to provide its customers with prototype molds for testing and evaluation. Using customer-supplied CAD data, the company used its CAM

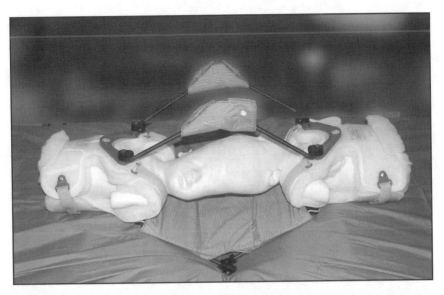

FIGURE A-10. One of the jacket designs is evaluated with the full-size skin model. *(Courtesy Medical Modeling, LLC)*

software to generate tool paths. Then Apex would cut P20 steel with its CNC equipment. This could be an intensive, lengthy process that often had to be repeated several times as customers discovered design modifications.

With its selective laser sintering system and LaserForm™ material (a 420 stainless steel material for tooling and metal parts), Apex has decreased the time for tooling to just one to four weeks, a reduction of 50–75%. Apex has also discovered that it can apply LaserForm, as shown in *Figure A-11*, to limited-run production molds with minimal surface finishing. For moderate tolerance requirements of 0.010 in. (0.25 mm), this rapid tooling solution has been applied to short-run production tooling, without secondary machining requirements.

Another advantage that Apex discovered is that it can incorporate features that would be difficult, if not impossible, to reproduce when machining. In one case, a core insert for a standard P20 tool was constructed in selective

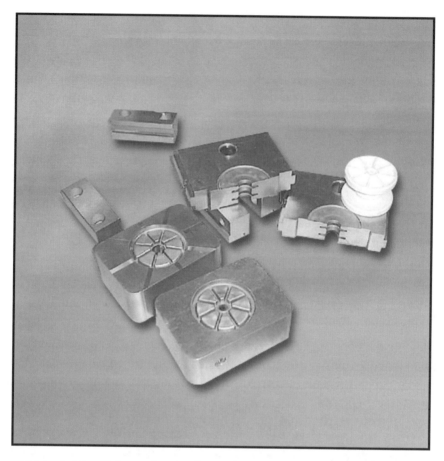

FIGURE A-11. Using selective laser sintering and 3D Systems' Laser-Form™, Apex Mold and Die has reduced delivery times by 50–75%. *(Courtesy 3D Systems)*

laser sintering. The insert incorporated very small, very long holes for the gas-assist method of injection molding.

Using LaserForm ST-100 tooling, Apex has successfully run over 35,000 cycles when molding 30% glass-filled nylon 6/12, without any appreciable wear.

RAPID PROTOTYPING CREDITED FOR MULTI-MILLION-DOLLAR CONTRACT

Technology: fused deposition modeling

Applications: form, fit, function, rapid manufacturing

Contributed by Stratasys, Inc. (Eden Prairie, Minnesota)

Bell & Howell's scanner division (Lincolnwood, Illinois) used rapid prototyping to customize scanner system components and win a multi-million-dollar contract. To win the contract, Bell & Howell had to design new scanner components, install working prototype parts, and quickly refine them to meet the customer's stringent specification milestones. The company credits fused deposition modeling as the tool that allowed them to achieve the goals and win the contract.

Bell & Howell's top-of-the-line scanner, the Copiscan® 8000 Plus series, lists for $39,000. This device scans 125 pages per minute, capturing data on both the front and back. Equally impressive is that the Copiscan can handle paper thicknesses ranging from heavy card stock to delicate rice paper—in the same scanning run.

Since no company offered a scanner capable of meeting the customer's specifications, several manufacturers were invited to compete for the order by designing a scanner that satisfied the demands.

Bell & Howell spent nearly five months customizing the scanner's mechanical function before it won the contract. During this process, numerous changes were made to components and each was prototyped using fused deposition modeling. After internal design reviews, the rapid prototypes were sent to the customer for trial. One area of focus was the document-feeder assembly, shown in *Figure A-12*. On average, Bell & Howell sent one feeder assembly per week. The customer would install the fused deposition modeling parts and provide feedback for design refinement.

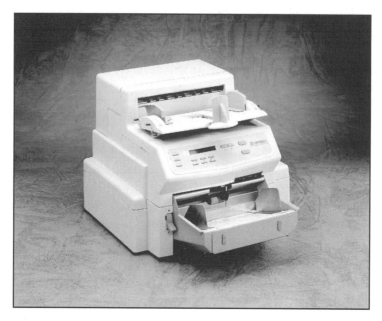

FIGURE A-12. The feeder assembly of the Copiscan® 8000 Plus by Bell & Howell was the focus of much of the design and prototyping effort. *(Courtesy Stratasys, Inc.)*

Regarding the feeder assembly modifications, Mike Scheller, director of mechanical engineering, stated, "We had four different ideas on how we might improve the feeder assembly. We were able to prototype them on the FDM® system, do some brief initial tests, and send the parts to the customer." *Figure A-13* shows the skimmer of the feed assembly.

According to Michael Jones, electro-mechanical technician, "You can have a 3D drawing with components that show movement, and it looks like it's functioning correctly. But when you build the solid model you see what you're missing. There were numerous times when we thought the design would work. We would build a prototype, install it in the scanner, and immediately catch a problem."

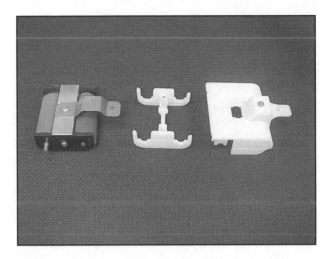

FIGURE A-13. The skimmer assembly of the document feeder showing fused deposition modeling prototypes (right, center). *(Courtesy Stratasys, Inc.)*

Prior to the purchase of its fused deposition modeling system, Bell & Howell's design cycle for new products was typically 18 months. This has been cut in half with an average of just nine months. Using the technology has also helped reduce tooling rework and its associated costs. Scheller says, "I'm much more confident that when we go to tooling, the part will work and look like it's supposed to. In the past, we spent a lot of money on tooling work that had to be redone."

Beyond form, fit, and functional prototypes, Bell & Howell also uses its FDM system for rapid manufacturing. Engineers found that the flag hold-down, when produced with fused deposition modeling, exceeded the performance requirements for a production unit (see *Figure A-14*). Rather than produce tooling for injection-molded flag hold-downs, the company produced over a thousand of them on the FDM system (see *Figure A-15*).

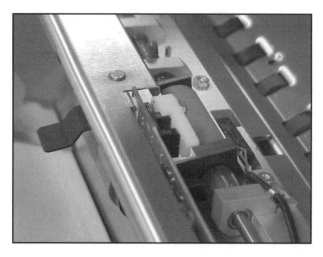

FIGURE A-14. The flag hold-down, in the center of the image, was rapid manufactured with fused deposition modeling. *(Courtesy Stratasys, Inc.)*

3D PRINTER APPLIED TO FUNCTIONAL TESTING AND RAPID MANUFACTURING

Technology: fused deposition modeling

Applications: functional testing, rapid manufacturing

Contributed by Stratasys, Inc. (Eden Prairie, Minnesota)

EOIR Technology (Occoquan, Virginia) was contracted to create a camera and mount for gun sights on M1 Abrams tanks and Bradley fighting vehicles, as shown in *Figure A-16*, for the Mississippi National Guard. When EOIR's subcontractor's designs for the mount failed performance tests, the company found itself up against a tight budget and contract deadline with no deliverable solution.

Using Solid Edge® CAD software and a Dimension™ 3D printer, EOIR was able to quickly evaluate models of alternative designs. The ABS parts proved tough enough to test directly on the armored vehicles without having to go through the costly, time-consuming process of machining them in aircraft-grade aluminum.

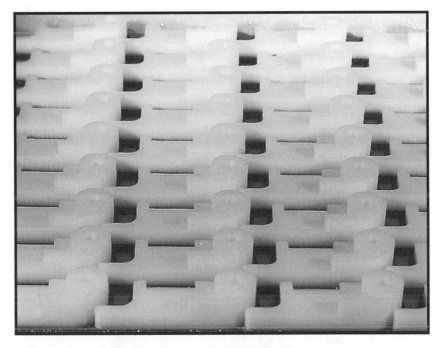

FIGURE A-15. A platform of rapid manufactured flag hold-downs. *(Courtesy Stratasys, Inc.)*

Functional models built with ABS plastic from the Dimension 3D printer proved so tough that EOIR manufactured the final mounts on the 3D printer itself (see *Figure A-17*). This saved time and dramatically reduced part cost. Using conventional means, manufacturing costs for these components would have exceeded $100,000. For less than $40,000, EOIR acquired a CAD software package, Dimension 3D printer, and ABS modeling materials, which allowed them to produce the 40 camera mounts internally. EOIR Technology project manager, John Moulton, notes, "If we had contracted with a separate machine shop to make these parts, not only would we not have made the schedule, but we wouldn't have a stay-behind piece of equipment that can continue to make money for us on future projects."

FIGURE A-16. The Bradley fighting vehicle was outfitted with rapid manufactured camera mounts (see also *Figure A-15*). *(Courtesy Stratasys, Inc.)*

CHEVY SSR RELIES ON RAPID PROTOTYPING

Technology: stereolithography

Applications: form, fit, patterns

Contributed by DSM Somos (New Castle, Delaware)

In Spring 2003, Chevrolet announced the much-anticipated Super Sport Roadster® (SSR). This stylized vehicle, which takes design queues from GM's trucks of the 1940s and 1950s, first appeared as a concept car at the North American International Auto show just two and a half years earlier. To meet an aggressive schedule while delivering a quality vehicle, the SSR counted on Auburn Engineering (Roches-

FIGURE A-17. Camera mount produced on the Dimension™ 3D printer. *(Courtesy Stratasys, Inc.)*

ter Hills, Michigan) to produce full-size interior and exterior prototypes. In turn, Auburn Engineering relied on DSM Somos® 9100 series resins for all of its stereolithography prototypes and patterns.

The SSR is part pickup and part roadster. The resulting design has created tremendous interest and a new automotive market that Chevrolet can call its own. In September 2000, GM made the decision to put the concept vehicle into production, knowing that a great deal was at stake with the launch of the SSR. The company hopes the SSR changes consumers' perceptions of GM and drives them to the dealers' showrooms.

The SSR's aim is to recapture GM's image as a bold manufacturer of quality vehicles. As a niche product with a forecast of just 11,000 units a year, the SSR is not intended to produce huge sales gains. As reported in the January 12, 2003 issue of *Lansing State Journal*, GM manufacturing spokesman Dan Flores said, "We're not going to regain five points of market share with the SSR, but we can erase perceptions of low quality and boring vehicles. That's what it's all about—creating the buzz that will hopefully be contagious and drive people to buy other GM products."

To maintain the high standards of quality and style, Auburn Engineering played a pivotal role in bringing the SSR to reality. In March 2002, Auburn Engineering was contracted to produce rapid prototypes of every plastic component in the SSR's interior. Instrument panel, dashboard, consoles, door panels, cup holders and stereo bezels were just the start of all the molded plastic parts prototyped. All of these parts were produced in Somos 9100 series resins because of the materials abilities to get the job done quickly and accurately.

To meet the aggressive delivery schedule of the SSR, the interior components went directly from design to production tooling. With GM's plans and hopes, there was no room for failure, problems, or delays. Therefore, after the release of the production tooling order, Auburn Engineering was asked to develop the prototypes for a full-size buck of the interior. Concurrent with the tooling order, designers would review the prototypes to confirm that form, fit, function, and flair were inherent in the SSR design.

Auburn Engineering selected Somos® 9120 resin for its combination of speed, durability, and flexibility. Assembly on the buck required snap fits and screw mountings. To achieve the desired results, the Somos 9120 prototypes had to withstand the rigors of assembly and the demands of the review process. Auburn Engineering believes that lesser resins would not have satisfied the application, and slower resins would have yielded a delay in the delivery of the prototypes.

"The Somos 9000 series resins rapidly created accurate, flexible parts for a complete replica of the vehicle's interior. This gave us the ability to create new parts or sections quickly when designs were changed," said Michael Vincek, national sales manager of Auburn Engineering. He further stated, "The Somos material properties allowed us to secure the parts on the buck without breaking and provided an accurate representation of the finished product, saving a significant amount of time and money in tooling."

Auburn Engineering had just two weeks to complete the prototypes. The last chance to modify any design for the production tooling would come from a design review meeting scheduled a few weeks after Auburn Engineering began receiving digital data. Using their five stereolithography systems and Somos 9120 resin, Auburn Engineering's team worked day and night to deliver the large quantity of prototypes on time. Evaluating form, fit, and visual appeal, each prototype was carefully crafted and finished for accuracy and aesthetics. Many of the prototypes were painted to simulate the color scheme and materials within the cockpit. Upon delivery, the prototypes were assembled onto the buck for the design review. Although tooling already had been released, and in some cases was in progress, the design team was able to detect and incorporate several modifications into the final production release.

Pleased with the earlier results of the Somos 9100 series prototypes for the interior buck, Chevrolet extended Auburn Engineering's work to exterior components. Prototypes included the fascia and the grille, claimed by Chevrolet to be one of the signature features of the SSR. For these applications, Auburn Engineering elected to use its rubber molding capabilities to produce urethane castings, as shown in *Figure A-18*. To fabricate the patterns for the large rubber molds, Auburn Engineering once again relied on the Somos 9100 series of resins.

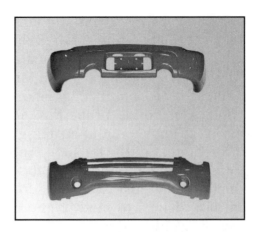

FIGURE A-18. Front and rear fascias for the Chevy SSR were cast from rubber molds and used on pre-production units. *(Courtesy Auburn Engineering)*

Patterns for rubber molding require a blend of properties, including excellent surface finish, tight dimensional accuracy, and durability. According to Vincek, "Patterns for rubber molds, especially those as large as the grille and fascia, must be rigid enough to withstand the weight and force of the rubber as it is being poured, yet durable enough to be extracted from the cured rubber." The combined strength, durability, and flexibility of the Somos 9100 series resins gave Auburn Engineering exactly what it needed.

Auburn Engineering believes its prototypes, the stereolithography models and the cast urethanes, were on SSR vehicles that made the auto show circuit over the past year. Although not confirmed, the company suspects the cover photo for the February 10, 2003 issue of *Business Week* offers a close-up of Auburn Engineering's work. Auburn Engineering takes pride in the fact that Chevrolet has relied on its prototypes to win over the automotive industry analysts, writers, dealers, and consumers.

The quality of the prototypes and of Auburn Engineering's work enabled the company to win the order for production tooling and molding on several under-the-hood components. According to William Carver, executive vice-president of Auburn Engineering, "It was a natural progression in the development cycle. Using our high-speed tooling and molding concept, we were able to move right into the production phase, which continued our role in saving a significant amount of time and money for the SSR program."

The creation of the SSR marks GM's new process to get fresh, exciting concepts from the computer to the auto shows and dealer showroom, and rapid prototyping plays an integral role. Like many companies in many industries, GM has experienced the advantages stereolithography and Somos resins offer when it comes to achieving rapid cycle times while delivering exceptional product quality.

The start of SSR's regular production began in Spring 2003. The limited-edition signature series was also available in early 2003. Chevrolet hopes the SSR changes consumer perceptions and gets them into the dealers' showrooms.

BIG RIGS BENEFIT FROM RAPID PROTOTYPING

Technology: selective laser sintering

Applications: form, fit, function, CAD verification

Contributed by Accelerated Technologies (Austin, Texas)

Beach Manufacturing Company (BMC) (Donnelsville, Ohio) specializes in the design and production of side-view mirror systems for tractor-trailers. Industry pressures prompted BMC to design a new mirror concept. Aerodynamic in shape and complementary in style and color, the "aero" mirror was a radical shift from traditional mirror designs.

To satisfy the demands of the OEMs and large carrier fleets, BMC had to quickly and economically produce the all-new B2000 aero mirror. Equally important, they had to meet these challenging goals without adding to their existing staff of one design engineer.

Initially, BMC relied on traditional model-making methods to produce mirror prototypes. However, with this process a single set of mirrors cost $30,000 and took more than a month for delivery. Realizing the old ways of doing business were too costly and time-consuming, BMC turned to Accelerated Technologies for help. With the company's selective laser sintering services, BMC received a full mirror assembly in just six days, costing just $7,850.

In all, BMC completed five phases of design and rapid prototyping. Throughout this process, the prototypes were used for functional testing and form/fit evaluations. The prototypes were also very effective as sales aids and communication tools.

Beyond the expected benefits of rapid prototyping, BMC found the process to be a valuable tool for verifying CAD databases. Two previously undetected flaws in the CAD file were discovered in the selective laser sintering models. With verification of the CAD files through rapid prototyping, there was a high degree of confidence in building production tools directly from CAD data without engineering drawings. Six injection molds were produced—without error or revision—in just nine weeks, a company record.

In all, BMC used selective laser sintering for 22 prototypes of six different components for the B2000 side-view mirror system. The total cost for all the models was just under $30,000. The total production time was only 31 days. Rapid prototyping allowed BMC to complete five design revisions in the same amount of time, and at the same cost, as one conventional iteration.

Overall, BMC reduced project cost by 25% and shaved six months from their time-to-market.

BIOMOLECULAR MODELS FOR RESEARCH AND EDUCATION

Technology: powder-binder printing

Application: presentation

Contributed by Milwaukee School of Engineering, Center for BioMolecular Modeling and 3D Molecular Designs

The Center for BioMolecular Modeling (CBM) at the Milwaukee School of Engineering (MSOE), in conjunction with 3D Molecular Designs, develops 3D physical models of proteins and other molecular structures using rapid prototyping technology. These models are physical representations based on the atomic coordinates of their structures as deposited in the protein data bank. Using five rapid prototyping technologies, including powder-binder printing from Z Corporation, the organizations offer 3D models for research and education purposes.

The complex models are incredibly useful as "thinking tools" in a research lab setting. They are ideally suited to small group discussions as portable, three-dimensional, graphical displays. The molecular models allow even the most experienced computer user to see things that are often elusive on the computer screen. When applied to a teaching environment, understanding is greatly improved for many students, because the molecular world is invisible and often complex. *Figures A-19* and *A-20* show two such models.

Rapid prototyping has generated models for many structures, including ribosome, nucleosome, beta-globin, HIV protease, and anthrax. The anthrax project is especially interesting since it was a result of collaboration between

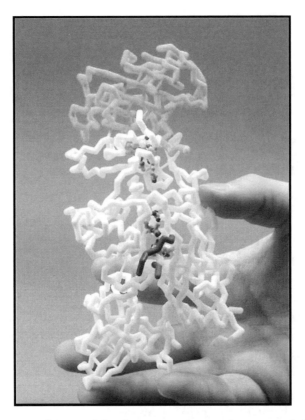

FIGURE A-19. This model of a protective antigen, an alpha-carbon backbone, was made with selective laser sintering. *(Courtesy Center for BioMolecular Modeling at the Milwaukee School of Engineering)*

CBM and Riverside University High School. The team, which was comprised of three high school seniors and Jeff Anderson, their teacher, joined with CBM to create physical models of anthrax proteins for use as presentation aids. The 3D models are used as a visualization tool that helps to explain the molecular mechanism of anthrax pathogenesis to students, teachers, and the general public.

A small start-up company, 3D Molecular Designs, LLC, was created with funds from a Phase I small business innovation research (SBIR) grant from the National Institute of

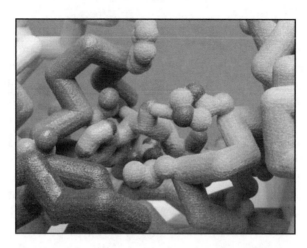

FIGURE A-20. This model of an edema factor active site was made of plaster using the color Z Corporation technology. *(Courtesy Center for BioMolecular Modeling at the Milwaukee School of Engineering)*

Health/National Center for Research Resources (NIH/NCRR) to Michael Patrick and Tim Herman, directors of CBM. The company was founded to commercialize the developments at CBM. Its services include the creation of custom models.

SKI BINDING INNOVATION FUELED BY RAPID PROTOTYPING

Technology: fused deposition modeling

Applications: form, fit, function

Contributed by Stratasys, Inc. (Eden Prairie, Minnesota)

Since 1927, Rotefella AS (Klokkarstua, Norway) has manufactured cross-country, telemark, and jumping ski bindings for consumers and the military. The company has a small work force but a big following, owning 50% of the binding market. Rotefella attributes product innovation to its leadership position. The company knows that moving swiftly from concept to market is more important than ever.

345

Having only a handful of design engineers in its small work force, Rotefella relies on outside collaborative partners Abry Industrial Design (Oslo, Norway) and The National Institute of Technology (TI) (Oslo, Norway).

Abry has noted several shifts in product development over the past decade. Early on, Abry helped its clients produce traditional models and prototypes fabricated from machined tooling. Inherent in this process were long lead times and the potential for high costs resulting from tooling rework. With the advent of 3D CAD, Abry shifted much of the design review and analysis away from prototypes to the virtual computer environment. While the CAD tools decreased time and cost, Abry found that designs lacked aesthetic value. With the emergence of RP, the company shifted its prototyping back to physical models while maintaining time and cost advantages.

With Abry's guidance, Rotefella now produces as many as 20 prototype iterations with fused deposition modeling prior to tooling. This contrasts to no models in the mid-90s and only two or three models in earlier years. Now part of every project, RP has reduced the product development cycle and enabled Rotefella to maintain its competitive advantage through design innovation.

Using the rapid prototyping resources of TI, namely the FDM 8000™, Rotefella reviews the ABS prototypes to discover mistakes and make changes early in the design cycle. With the speed of the rapid prototypes and the ability to discover changes early, Abry finds that the typical time reduction for design projects is 8–12 months. Equally significant is the improvement in product quality. With the fused deposition modeling prototypes, the company can identify the need for changes much earlier in the project. Having the physical models offers reduced product development time, better designs, less expensive tooling, and lower risk.

Even with the obvious advantages, not everyone immediately embraced rapid prototyping for the project. Rotefella's toolmakers were concerned about becoming obsolete. However, after working with the fused deposition modeling prototypes, they have become strong supporters. The rapid prototypes allow the toolmakers to review early iterations of each model and offer design input prior to cutting tooling. This collaboration reduces last minute rework, which makes the operation less stressful and less costly. With this success, the use of the rapid prototypes has progressed farther downstream in the manufacturing process. Rotefella now uses rapid prototypes as training aids for the assemblers and as fixtures for assembly.

PRO SERIES™ POWER FASTENING TOOL PERFECTED WITH RAPID PROTOTYPING

Technologies: stereolithography, selective laser sintering, plaster mold casting

Applications: form, fit, function, patterns, sales promotion

Contributed by Accelerated Technologies (Austin, Texas)

Senco Products (Cincinnati, Ohio) uses rapid prototyping for multiple applications. For its pneumatic nailers and staplers, Senco relied on Accelerated Technologies' rapid prototyping services for product announcements, engineering reviews, styling reviews, functional analysis, and much more. During the development of the PRO Series™ line for the do-it-yourself market, Senco applied rapid prototyping to its new FinishPro™ 35 finish nailer.

Over the past 50 years, Senco has designed and manufactured power-fastening products with the right blend of quality, performance, and price. While these are critical factors, the company understands that appearance and feel are equally important because they convey quality and attract consumer attention. As a result, Senco incorporated a new, consistent look and feel across its line of PRO Series products. With Accelerated Technologies' selective laser sintering and stereolithography services and the application of plaster mold casting, Senco designed the FinishPro 35 with the qualities of a professional power tool priced for the do-it-yourself market.

For the first phase of rapid prototyping, Senco requested selective laser sintering models of the outer housing of the FinishPro 35. Early in the design process, Senco's goal was to evaluate form, fit, and ergonomics. In just five days, Accelerated Technologies delivered the housings. In the review, Senco uncovered a few minor modifications that needed to be made to the internal structure. More importantly, the company detected exterior design flaws that detracted from the visual appeal and consistency of design style for the product line. Jeff Koonce, engineer at Senco Products, stated, "In CAD, it is just not possible to find all of these changes. It takes a physical model for the engineering team to see them, especially when it comes to the aesthetic design."

Following the engineering review, Senco "trade dressed" the selective laser sintering models with the new look of the PRO Series line and used them to introduce the product to its sales team (see *Figure A-21*). Painted like the final product and weighted to demonstrate balance and ergonomics, the models gave the sales team its first look at the FinishPro 35. The prototypes helped the sales team to understand and appreciate the new product, including a hands-on feel for the weight and balance of the new tool. Senco believes the rapid prototypes heightened enthusiasm and understanding, which in turn helped the sales force promote the new line to their customers well before any product was available. According to Koonce, "The selective laser sintering prototypes were an enormous benefit. Showing off the FinishPro 35 and the new Senco look

FIGURE A-21. A trade-dressed selective laser sintering model of the FinishPro™ 35 nailer. *(Courtesy Accelerated Technologies)*

jump-started our sales efforts." The dressed models were also presented in early promotional materials for the FinishPro 35.

After incorporating design changes from the engineering review, Senco needed functional prototypes that could withstand the internal pressures required for firing a finishing nail. An area of great importance to the efficiency and power of the FinishPro 35 is the design of the firing and return chambers. Senco has followed established design rules over the past 50 years. The company knows that small changes in the pressurized chamber can have a big impact on the tool's performance. So Senco once again turned to Accelerated Technologies. Wanting to test all aspects of the design, Senco requested metal prototypes. To save time and money—compared to CNC machining and die casting—Accelerated Technologies recommended plaster mold casting, a process unfamiliar to Senco. Using stereolithography for pattern generation, five sets of cast aluminum housings were delivered in just four weeks. Within days, Senco had the fully functional prototypes firing a variety of brads and finishing nails. Senco was pleased to find that the design, and the prototypes, performed flawlessly.

Jeff Koonce commented, "The rapid prototypes were well worth the investment—they helped the design team identify the need for changes that resulted in a better product."

The design work and rapid prototypes have paid off. The FinishPro 35 has exceeded Senco's sales expectations. To see the quality of work and the results that rapid prototyping can produce, visit Lowe's home improvement centers, one of Senco's largest retail channels for the FinishPro 35.

Appendix B

Additional Resources

The following resource listing is also available online at www.tagrimm.com/book/resources.html.

CONTRIBUTORS

The author would like to thank and acknowledge the following organizations that contributed to *User's Guide to Rapid Prototyping*.

3D Molecular Designs
  (www.3dmoleculardesigns.com)
  2223 N. 72nd St., Wauwatosa, WI 53213, USA
  Custom molecular models with rapid prototyping

3D Systems, Inc.
  (www.3dsystems.com)
  26081 Avenue Hall, Valencia, CA 91355, USA
  Rapid prototyping equipment and materials

Accelerated Technologies
  (www.acceleratedtechnologies.com)
  1611 Headway Circle, Building 1,
  Austin, TX 78754, USA
  Rapid prototyping, design and tooling services, and contract manufacturing

Armstrong Mold
(www.armstrongmold.com)
6910 Manlius Center Rd., E. Syracuse, NY 13057, USA
Prototyping and low-volume production of metal and plastic parts

Auburn Engineering
(www.auburn.com)
2961 Bond St., Rochester Hills, MI 48309, USA
Design, prototyping, and tooling services

BASTECH, Inc.
(www.bastech.com)
3931 Image Dr., Dayton, OH 45414, USA
Engineering, rapid prototyping, and rapid tooling services

Center for BioMolecular Modeling at the Milwaukee School of Engineering (MSOE)
(www.rpc.msoe.edu)
Rapid Prototyping Center, 1025 N. Broadway St., Milwaukee, WI 53202, USA
Molecular models for research and education

Clemson University Center for Advanced Manufacturing
(www.vr.clemson.edu/rp/index.htm)
206 Fluor Daniel EIB, Clemson, SC 29634, USA
Rapid prototyping research and consortium

D-M-E MoldFusion
(www.dme.net/wwwdme/moldfusion.asp)
29111 Stephenson Hwy.,
Madison Heights, MI 48071, USA
Rapid tooling services and conformal cooling research

DSM Somos
(www.dsmsomos.com)
2 Penn's Way, Suite 401, New Castle, DE 19720, USA
Rapid prototyping materials for stereolithography and other technologies

Dynacept Corporation
(www.dynacept.com)
2 International Blvd., Brewster, NY 10509, USA
*Rapid prototyping, model making, tooling, and molding
services*

Extrude Hone
(www.prometal-rt.com)
1 Industry Blvd., P.O. Box 1000, Irwin, PA 15642, USA
Rapid tooling equipment

Fisher Design
(www.fisherdesign.com)
2830 Victory Parkway, Suite 300,
Cincinnati, OH 45206, USA
*Design, prototyping, product development, and
marketing services*

General Pattern Company
(www.generalpattern.com)
3075 84th Lane N.E., Blaine, MN 55449, USA
*Rapid product development services including rapid
prototyping, tooling, and molding for plastic and metal*

Leyshon Miller Industries
(www.lmidesign.com)
534 N. 1st St., Cambridge, OH 43725, USA
*Product development services including design,
prototyping, tooling, and manufacturing*

LGM (Laser Graphic Manufacturing)
(www.lgmmodel.com)
23698 U.S. Highway 24, Minturn, CO 81645, USA
Architectural modeling services and software solutions

Materialise GmbH
(www.materialise.com)
6111 Jackson Rd., Ann Arbor, MI 48103, USA
Headquarters: Leuven, Belgium
*Software for rapid prototyping, rapid tooling, and
reverse engineering*

Medical Modeling, LLC
(www.medicalmodeling.com)
17301 W. Colfax Ave., Suite 300,
Golden, CO 80401, USA
Medical models from CT and MRI data for surgical review and planning

Ralph S. Alberts Company, Inc.
(www.rsalberts.com)
60 Choate Circle, Montoursville, PA 17754, USA
Custom molder and moldmaker with specialization in epoxy, urethane, and silicone rubber molds

Rapid Prototyping Association
(www.sme.org/rpa)
Society of Manufacturing Engineers
One SME Drive, P.O. Box 930, Dearborn,
MI 48121, USA
Sponsor of Rapid Prototyping & Manufacturing conference

Rapid Prototyping Report
(www.cadcamnet.com)
CAD/CAM Publishing, Inc.
1010 Turquoise St., Suite 320,
San Diego, CA 92109, USA
Monthly publication on rapid prototyping, tooling, and manufacturing

Roland DGA
(www.rolanddga.com)
15363 Barranca Parkway, Irvine, CA 92618, USA
Rapid prototyping equipment, including milling and scanning systems

Solidica, Inc.
(www.solidica.com)
3941 Research Park Dr., Suite C,
Ann Arbor, MI 48108, USA
Rapid tooling equipment for metal molds and dies

Stratasys, Inc.
(www.stratasys.com)
14950 Martin Dr., Eden Prairie, MN 55344, USA
Rapid prototyping equipment

Ultimate Solutions, Inc.
(www.ultimatesolutions-inc.com)
13011 Bristol Berry Dr., Cypress, TX 77429, USA
Rubber molding services and training

University of Louisville, Rapid Prototyping Center
(www.louisville.edu/speed/rpc)
Vogt Building, Room 101, Louisville, KY 40292, USA
Rapid prototyping and tooling research and consortium

University of Stellenbosch
(www.ie.sun.ac.za)
Department of Industrial Engineering, Private Bag X1,
Matieland 7602, South Africa
Rapid prototyping, tooling and manufacturing research

Wohlers Associates, Inc.
(www.wohlersassociates.com)
Oak Ridge Business Park, 1511 River Oak Dr.,
Fort Collins, CO 80525, USA
Consulting for rapid prototyping, tooling and manufacturing

Worldwide Guide to Rapid Prototyping
(home.att.net/~castleisland)
19 Pondview Road, Arlington, MA 02474, USA
Comprehensive web resource with extensive resource lists and technology information

Z Corporation
(www.zcorp.com)
20 North Ave., Burlington, MA 01803, USA
Rapid prototyping equipment

MANUFACTURERS (RAPID PROTOTYPING HARDWARE, SOFTWARE, AND MATERIALS)

Equipment

3D Systems, Inc. (www.3dsystems.com)
26081 Avenue Hall, Valencia, CA 91355, USA

AeroMet Corporation (www.aerometcorp.com)
7623 Anagram Dr., Eden Prairie, MN 55344, USA

Arcam AB (www.arcam.com)
Krokslätts Fabriker 30, SE-431 37 Mölndal, Sweden

Autostrade Co., Ltd. (www.autostrade.co.jp)
13-54 Ueno-machi, Oita-City, Oita 870-0832, Japan

CMET (www.cmet.co.jp)
Japan

Concept Laser, GmbH (www.concept-laser.de)
An der Zeil 8, 96215 Lichtenfels, Germany

Cubic Technologies (www.cubictechnologies.com)
1000 E. Dominguez St., Carson, CA 90746-3608, USA

Denken Engineering (www.coara.or.jp/~dkslp/my.html)
Japan

Envisiontec, GmbH (www.envisiontec.de)
Elbestrasse 10, D-45768 Marl, Germany

EOS GmbH (www.eos-gmbh.de)
Robert-Stirling-Ring 1,
D-82152 Krailling/Munich, Germany

Extrude Hone (www.prometal-rt.com)
1 Industry Blvd., P.O. Box 1000, Irwin, PA 15642, USA

Fockele und Schwarze (www.fockeleundschwarze.de)
Alter Kirchweg 34,W-33178 Borchen-Alfen, Germany

Generis GmbH (www.generis.de)
Am Mittleren Moos 15,
D-86167 Augsburg, Germany

Kinergy Pte., Ltd. (www.kinergy.com.sg)
Block 5002, #03-08, Techplace II, Ang Mo Kio Ave. 5,
S 569871, Singapore

Kira Corporation
(www.kiracorp.co.jp/EG/pro/rp/top.html)
Tomiyoshishinden, Kira-cho, Hazu-gun, Aichi Pref., Japan

Meiko (www.meiko-inc.co.jp)
Japan

Objet Geometries, Ltd. (www.2objet.com)
2 Holzman St., Science Park, P.O. Box 2496,
Rehovot 76124, Israel

Optomec, Inc. (www.optomec.com)
3911 Singer Blvd., N.E., Albuquerque, NM 87109, USA

Phenix Systems (www.phenix-systems.com)
ZAC du Brezet EST, 29, rue Georges Besse,
63100 Clermont-Ferrand, France

POM Group, Inc. (www.pom.net)
Advanced Product Development Center, 2350 Pontiac Rd.,
Auburn Hills, MI 48326, USA

RSP Tooling, LLC (www.rsptooling.com)
30700 Carter Rd., Solon, OH 44139, USA

Sanders Design International, Inc.
(www.sandersdesign.com)
37 Wilton Rd., Milford, NH 03055, USA

Solidimension, Ltd. (www.solidimension.com)
Shraga Katz Building, Be'erot Itzhak, 60905, Israel

Solidscape, Inc. (www.solid-scape.com)
316 Daniel Webster Hwy., Merrimack, NH 03054, USA

Solidica, Inc. (www.solidica.com)
3941 Research Park Dr., Suite C, Ann Arbor, MI 48108, USA

Sony Precision Technology, Inc. (www.sonypt.com)
20381 Hermana Circle, Lake Forest, CA 92630, USA

Speed Part AB (www.speedpart.se)
Rådanäs, 435 33 Mölnlycke, Sweden

Stratasys, Inc. (www.stratasys.com)
14950 Martin Dr., Eden Prairie, MN 55344, USA
Therics, Inc. (www.therics.com)
115 Campus Dr., Princeton, NJ 08540, USA
Unirapid (www.unirapid.com)
Z Corporation (www.zcorp.com)
20 North Ave., Burlington, MA 01803, USA

Materials

Axson NA (www.axson.com)
1611 Hults Dr., Eaton Rapids, MI 48827, USA
BJB Enterprises, Inc. (www.bjbenterprises.com)
14791 Franklin Ave., Tustin, CA 92780, USA
Bolson Materials Corporation (www.bolsonmaterials.com)
10612-105 Avenue, Edmonton, Alberta T5H 0L2, Canada
DSM Somos (www.dsmsomos.com)
2 Penn's Way, Suite 401, New Castle, DE 19720, USA
Freeman Manufacturing & Supply Company
(www.freemansupply.com)
1101 Moore Rd., Avon, OH 44011, USA
Innovative Polymers
(www.thomasregister.com/olc/innovative-polymers)
208 Kuntz St., St. Johns, MI 48879, USA
Sibco, Inc. (www.sibcoinc.com)
1100 Hilton St., Ferndale, MI 48220, USA
Tuxedo Photopolymers, American Dye Source, Inc.
(www.tuxedoresin.com)
555 Morgan Blvd., Baie D'Urfe, Quebec H9X 3T6, Canada
Vantico, Inc., Renshape Solutions
(www.renshape.com)
4917 Dawn Ave., East Lansing, MI 48823, USA

Repair, Parts, and Service

Laser Innovations (www.solidimaging.com)
668 Flinn Ave., Suite #22, Moorpark, CA 93021, USA
National RP Support (www.rpsupport.com)
P.O. Box 276, Pella, IA 50219, USA
Sibco, Inc. (www.sibcoinc.com)
1100 Hilton St., Ferndale, MI 48220, USA

Software

Actify, Inc. (www.actify.com)
  60 Spear St., 5th Floor, San Francisco, CA 94105, USA
Brock Rooney & Associates, Inc.
  915 Westwood, Birmingham, MI 48009, USA
Delcam, plc (www.delcam.com)
  3270 Electricity Dr., Windsor, Ontario N8W 5J1, Canada
  Headquarters: Birmingham, UK
Delft Spline Systems
  (www.deskproto.com)
  Vogelsanglaan 30, 3571 ZM Utrecht, Netherlands
Deskartes Oy
  (www.deskartes.com)
  Pihlajatie 28, Fin-00270, Helsinki, Finland
Floating Point Solutions (www.fpsols.com)
  L-54, Housing Board Colony, Alto Porvorim, Goa, India
Inus Technology, Inc. (www.inustech.com)
  3003 N. First St., San Jose, CA 95134, USA
  Headquarters: Seoul, Korea
Marcam Engineering, GmbH
  (www.marcam.de)
  Fahrenheitstrasse 1, D-28359 Bremen, Germany
Materialise, GmbH (www.materialise.com)
  6111 Jackson Rd., Ann Arbor, MI 48103, USA
  Headquarters: Leuven, Belgium

Metris N.V.
(www.paraform.com)
Interleuvenlaan 15D, B-3001 Leuven, Belgium
Raindrop Geomagic, Inc. (www.geomagic.com)
P.O. Box 12219, Research Triangle Park, NC 27709, USA
Solid Concepts, Inc. (www.solidconcepts.com)
28231 Avenue Crocker, Bldg. 10, Valencia, CA 91355, USA

Consulting and Training

Edward Mackenzie, Ltd. (www.edwardmackenzie.com)
Baileycroft House, Chapel Lane, Wirksworth, Derbyshire
DE4 4FF, UK
Ennex Corporation (www.ennex.com)
Santa Barbara, CA, USA
New Product Dynamics (www.newproductdynamics.com)
3493 N.W. Thurman St., Portland, OR 97210, USA
T. A. Grimm & Associates, Inc. (www.tagrimm.com)
3028 Beth Court, Edgewood, KY 41017, USA
Warner Technologies
4123 Oceana, Waterford, MI 48328, USA
Wohlers Associates, Inc. (www.wohlersassociates.com)
OakRidge Business Park, 1511 River Oak Dr., Fort Collins,
CO 80525, USA

Publications

Books

Burns, Marshall. 1993. *Automated Fabrication: Improving Productivity in Manufacturing.* Englewood Cliffs, NJ: Prentice Hall.

Chua, C. K., Leong, K.F., and Lim, C.S. 2003. *Rapid Prototyping; Principles and Applications.* River Edge, NJ: World Scientific Publishing Co.

Cooper, Kenneth G. 2001. *Rapid Prototyping Technology: Selection and Application.* New York: Marcel Dekker, Inc.

Gibson, Ian, ed. 2002. *Software Solutions for Rapid Prototyping.*

Hilton, Peter and Jacobs, Paul, eds. 2000. *Rapid Tooling: Technologies and Industrial Applications.* New York: Marcel Dekker, Inc.,

Jacobs, Paul F. 1992. *Rapid Prototyping & Manufacturing: Fundamentals of Stereolithography.* Dearborn, MI: Society of Manufacturing Engineers.

Jacobs, Paul F. 1995. *Stereolithography and Other RP&M Technologies: From Rapid Prototyping to Rapid Tooling.* Dearborn, MI: Society of Manufacturing Engineers.

Leu, Donald. 2000. *Handbook of Rapid Prototyping and Layered Manufacturing.* New York: Elsevier/Academic Press.

McDonald, J. A., Ryall, C. J., and Wimpenny, D. I., eds. 2001. *Rapid Prototyping Casebook.* London, England: Professional Engineering Publications Ltd.

Pham, D. T. and Dimov, S. S. 2001. *Rapid Manufacturing: The Technologies and Applications of Rapid Prototyping and Rapid Tooling.* New York: Springer Verlag Publishing.

Wohlers, Terry. 2003. *Wohlers Report; Rapid Prototyping & Tooling State of the Industry.* Fort Collins, CO: Wohlers Associates, Inc. (www.wohlersassociates.com)

Periodicals and Journals

Rapid Prototyping Journal. Bradford, UK: EmeraldJournals(www.emeraldinsight.com/rpsv/rpj.htm)

Rapid Prototyping Report. San Diego, CA: CAD/CAM Publishing, Inc. (www.cadcamnet.com)

Time Compression Technologies. Doylestown, PA:
  Communication Technologies, Inc.
  (www.timecompress.com)

Time Compression Technologies. Chester,UK:
  Rapid News Publications, plc
  (www.time-compression.com)

EVENTS

Euromold, Frankfurt, Germany
  (217.7.92.74/english/index.php4)
  Demat, GmbH, Poststraße 2-4,
  60329 Frankfurt/Main, Germany
Rapid Prototyping & Manufacturing (www.sme.org/rapid)
  Society of Manufacturing Engineers, One SME Dr.,
  P.O. Box 930, Dearborn, MI 48121, USA
Siggraph (www.siggraph.org)
  ACM Siggraph Conference Management
  401 N. Michigan Ave.,
  Chicago, IL 60611, USA
Solid Freeform Fabrication
  (utwired.engr.utexas.edu/lff/symposium/index.cfm)
  University of Texas, Mechanical Engineering, MC C2200,
  204 East Dean Keeton St., Austin, TX 78705, USA
TCT Exhibition (www.time-compression.com)
  Rapid News Publications, plc, Protel House,
  10 Hunters Walk, Canal St., Chester CH1 4EB, UK

INFORMATION

Websites

University of Utah
  (www.cc.utah.edu/~asn8200/rapid.html)

Wohlers Associates, Inc.
  (www.wohlersassociates.com)
Worldwide Guide to Rapid Prototyping, Castle Island
  (home.att.net/~castleisland)

Forums and Mailing Lists

Rapid Prototyping Mailing List (rapid.lpt.fi/rp-ml)

ASSOCIATIONS

Association for RP Companies in the Netherlands

Association of Professional Model Makers
  (www.modelmakers.org)
  6502 Shiner St., Austin, TX 78729, USA

Australia's QMI Solutions, Ltd.
  (www.qmisolutions.com.au)
  Miles Platting & Logan Roads,
  Eight Mile Plains QLD 4113, Australia

Canadian Association of Rapid Prototyping, Tooling,
  and Manufacturing (www.nrc.ca/imti)

Chinese Rapid Forming Technology Committee
  (www.geocities.com/CollegePark/Lab/8600/rftc.htm)

Danish Technological Institute

Finnish Rapid Prototyping Association
  (www.isv.hut.fi/firpa)

French Rapid Prototyping Association
  (www.art-of-design.com/afpr)

Germany's NC Society (www.ncg.de)

Global Alliance of Rapid Prototyping Associations
  (GARPA) (www.garpa.org)

Hong Kong Society for Rapid Prototyping Tooling
  and Manufacturing
  (hkumea.hku.hk/~CRPDT/RP&T.html)

Italian Rapid Prototyping Association (www.apri-rapid.it)
Viale Fulvio Testi, 128-20092 Cinisello Balsamo (MI), Italy

Japanese Association of Rapid Prototyping Industry
(www.rpjp.or.jp)

Rapid Product Development Association of South Africa

Rapid Prototyping Association (www.sme.org/rpa)
Society of Manufacturing Engineers, One SME Drive,
P.O. Box 930, Dearborn, MI 48121, USA

Swedish Industrial Network on FFF
(www.ivf.se/FFF/fffblad.pdf)
IVF, Brinellvägen 68, 100 44 Stockholm, Sweden

UK's Rapid Prototyping and Manufacturing Association
(www.imeche.org.uk/manufacturing/rpma)

RESEARCH AND EDUCATION

Center for BioMolecular Modeling at the Milwaukee School
of Engineering (MSOE) (www.rpc.msoe.edu)
Rapid Prototyping Center, 1025 N. Broadway St.,
Milwaukee, WI 53202, USA

Center for Rapid Product Development
(soe.unn.ac.uk/crpd/index.htm)
School of Engineering and Technology
Northumbria University, Ellison Place,
Newcastle upon Tyne NE1 8ST, UK

Clemson University Center for Advanced Manufacturing
(www.vr.clemson.edu/rp/index.htm)
206 Fluor Daniel EIB, Clemson, SC 29634, USA

Institute for Rapid Product Development,
University of Applied Science
St. Gallen (FHS) (www.fhsg.ch/rpd)
Tellstrasse 2, Postfach 664,
CH-9001 St. Gallen, Switzerland

Institute of Technology, Tallaght
(www.it-tallaght.ie)
Department of Mechanical Engineering
Tallaght, Dublin 24, Ireland

Massachusetts Institute of Technology,
Laboratory for Manufacturing and Productivity
(web.mit.edu/lmp/www)
77 Massachusetts Ave., Cambridge, MA 02139, USA

Milwaukee School of Engineering,
Rapid Prototyping Center
(www.rpc.msoe.edu)
1025 N. Broadway St., Milwaukee, WI 53202, USA

Ngee Ann Polytechnic (www.np.edu.sg)
5335 Clementi Rd., Singapore 599489

Stanford University, Rapid Prototyping Lab
(www-rpl.stanford.edu)
Building 530, Room 226, Stanford, CA 94305, USA

Rapid Prototyping and Manufacturing Institute
(www.rpmi.marc.gatech.edu)
Manufacturing Research Center
Georgia Institute of Technology, 813 Ferst Dr., N.W.,
Atlanta, GA 30332, USA

Tyler School of Art
(www.temple.edu/crafts)
Temple University, 7725 Penrose Ave.,
Elkins Park, PA 19027, USA

University of California-Irvine
(www.eng.uci.edu/%7emelissao/droplab/droplab.htm)
Department of Mechanical and Aerospace Engineering
Irvine, CA 92697, USA

University of Louisville, Rapid Prototyping Center
(www.louisville.edu/speed/rpc)
Vogt Building, Room 101, Louisville, KY 40292, USA

University of Texas
(utwired.engr.utexas.edu/lff)
University of Texas, Mechanical Engineering, MC C2200,
204 E. Dean Keeton St., Austin, TX 78705, USA

University of Stellenbosch
(www.ie.sun.ac.za)
Department of Industrial Engineering, Private Bag X1,
Matieland 7602, South Africa

University of Warwick Rapid Prototyping & Tooling Center
(www.warwick.ac.uk/atc/rpt/)
Advanced Technology Center
Warwick Manufacturing Group, Coventry CV4 7AL, UK

APPENDIX C

Glossary

Portions of this glossary have been adapted from the *RP Glossary* published by the Rapid Prototyping Association of the Society of Manufacturing Engineers (RPA/SME) in 2003.

2D
Abbreviation for two-dimensional. Often applied to the description of CAD systems (for example, 2D CAD), the term indicates the resulting file is a flat representation with dimensions in only the X and Y axes.

3D
Abbreviation for three-dimensional. Often applied to the description of CAD systems (for example, 3D CAD), the term indicates the resulting file is a volumetric representation with dimensions in the X, Y, and Z axes.

3DP
See 3D printing.

3D Keltool®
An indirect rapid tooling process where powdered metals are formed against a pattern and sintered. This technology is owned and licensed by 3D Systems.

3D printing
1) Rapid prototyping processes using systems that are low cost, small in size, fast, easy to use, and often suitable for an office environment. Original process and terminology developed at Massachusetts Institute of Technology (MIT); now commonly used as a generic term.

2) Collective term for all rapid prototyping activities.

3 axis

Devices that have simultaneous motion in the *X*, *Y*, and *Z* axes.

5 axis

Devices that have simultaneous motion in the *X*, *Y*, and *Z* axes and two rotational axes.

A

ACES™

Acronym for accurate clear epoxy solid. A stereolithography build style that offered increased accuracy and improved surface finish, when compared to earlier build styles.

additive manufacturing

See rapid prototyping or rapid manufacturing.

alpha test

In-house testing of preproduction products to find and eliminate the most obvious design deficiencies. *See also beta test.*

ARP

Acronym for additive rapid prototyping. *See rapid prototyping.*

ASCII

Acronym for American Standard Code for Information Interchange. Coding system for representing characters in numeric form. A text file is created that can be displayed on a screen or printed without special formatting or specific software program requirements.

aspect ratio

Relative relationship between height and width. Expressed in integer form (not percentage) as the ratio of height to width, where each is divided by the width to yield a ratio of *X*:1.

associative geometry

Placing and controlling graphic elements based on their relationship to previously defined graphic elements. Elements placed associatively maintain their relationship as an element is manipulated.

associativity

Operating under a single, integrated database structure. Allows changes in any application (for example, design, drawing, assembly, etc.) to be reflected instantly in all associated applications as well as in every deliverable (for example, drawings, bills-of-materials, and numerical control [NC] tool paths).

axis (CAD)

Imaginary line segment upon which all measurements are made when creating or documenting a computer-aided design (CAD) model in 3D space. The complete Cartesian coordinate system comprises the X, Y, and Z axes.

B

BASS™

Acronym for break-away support structure. A style of support structure for the fused deposition modeling process, which is manually removed after prototype creation.

benching

For shop-floor or model-making operations, the process of finishing a part or prototype, typically with manual operations and hand tools, for example, sanding, filing, joining, and bonding.

b-rep

Abbreviation for boundary representation. Computer-aided design (CAD) software methodology defining the model as a set of vertices, edges, and faces (points, lines, curves, and surfaces).

b-spline

Abbreviation for bi-cubic spline. Sequence of parametric polynomial curves (typically quadratic or cubic polynomials) forming a smooth fit between a sequence of points in 3D space.

beta test

External operation of pre-production products in field situations to find those faults that go undetected in controlled in-house tests, but may occur when in actual use. *See also alpha test.*

bezier curves

Quadratic (or greater) polynomial for describing complex curves and surfaces.

binary system

Numbering system in base two, using ones and zeros.

bit

Single-digit number in base two, or binary notation (either a 1 or 0). The smallest piece of information understood by a computer.

bitmap

Matrix of pixels representing an image.

blow molding

Manufacturing process in which plastic material, in a molten state, is forced under high pressure into a mold, causing the plastic to conform to the shape of the tool with a consistent wall thickness. The process is often used to produce hollow items such as bottles.

BOM

Acronym for bill of materials. Listing of required quantities of all subassemblies, intermediate parts, and raw materials going into a parent assembly.

BPM

Acronym for ballistic particle manufacturing. Rapid prototyping process where wax materials are deposited with a multi-axis, inkjet print head. This process is no longer available.

bridge tooling

Relating to molds or dies intended to fill the demand between early prototype, or soft tooling, and production tooling.

build time

Length of time for the physical construction of a rapid prototype, excluding preparation and post-processing time. Also known as run time.

C

CAD

Acronym for computer-aided design or computer-aided drafting. A software program for the design and documentation of products in either two or three-dimensional space.

CAE

Acronym for computer-aided engineering. A software program that uses computer-aided design (CAD) data for the analysis of mechanical and thermal attributes and behavior. This is accomplished through the use of finite element analysis (FEA) software.

CAM

Acronym for computer-aided manufacturing. A software program that uses the design data of computer-aided design (CAD) to build tool paths and similar manufacturing data for the purposes of machining prototypes, parts, fixtures, or tooling.

cavity

Mold component that forms the exterior surface of the closure.

children

1) Components of a design instance in a product structure tree. Also referred to as parts.

2) Nodes in a database tree structure with a parent.

3) Features in parametric modeling that are dependent

on others for establishing location in space. If the parent features are changed drastically, the children can become "orphans," or unassociated.

chord

A line segment that connects two distinct points on an arc.

chord height

Distance from the chord to the surface that the chord approximates. One of several terms that relate to the control and tolerance of the stereolithography (STL) file.

CMM

Acronym for coordinate measuring machine. A device used to determine 3D spatial coordinates from a physical part. The output is typically used for inspection and can be used for reverse engineering.

CNC

Acronym for computer numerical control. Numerical control (NC) system in which the data handling, control sequences, and response to input are determined by an on-board computer system at the machine tool.

coincidence

Geometry that occupies the same spatial location. For example, coincident lines can have differing lengths while one occupies the same location as the other.

compression

Process of compacting digital data to reduce file size for electronic transmission or data archiving.

computer model

Set of computer data representing a product or process and capable of being used to simulate the physical product or process behavior.

concept model

Physical model intended primarily for design review and not meant to be sufficiently accurate or durable for full functional or physical testing. Examples: foam models, 3D printed parts, rapid prototype parts.

concept optimization/concept study

Research approach that evaluates how specific product benefits or features contribute to a concept's overall appeal to consumers. Included are product development tasks that help determine unknowns about the market, technology, or production processes.

concurrent engineering

Organization of product design, development, production planning, and procurement processes, which occurs in parallel rather than in series. Includes the use of a project-oriented team to get input from all concerned parties.

conformal cooling

Water lines in tooling that follow the geometry of the part to be produced, providing higher cooling rates and lower cycle times.

core

Mold component that forms the internal surface of the closure.

CSG

Acronym for constructive solids geometry. A computer-aided design (CAD) modeling technique that uses a hierarchical representation of instances of solids and combination operations (union, intersection, difference).

CT

Acronym for computer-aided tomography.
1) Scanning system based on X-ray technology used to reverse engineer or dimensionally verify physical parts.
2) X-ray based volumetric scanning used for solid objects (for example, bone in humans, but also industrial components) with internal features.

cycle time

Period between the start of an operation and the start of the next occurrence of the same operation.

D

D_p

Penetration depth. A variable for photo-curable materials that specifies the depth of solidification at a known level of power input. Combined with E_c, (critical energy), the two identify the photo speed of a resin.

design for manufacturability

Process to insure that a product or its components can be manufactured. The objective is to maximize the process rate and minimize the cost to produce.

DFA

Acronym for design for assembly. Application of a design philosophy to insure parts and part designs are optimized for the assembly process. This step is important for assurance that parts can be handled, oriented, and positioned accurately.

DFM

Acronym for design for manufacturability. *See design for manufacturability.*

die casting

Manufacturing process that produces metal components through the pressurized injection of molten alloys into a metal tool (die). The process is typically used for high-volume production.

digital modeling

The concept of holding the master product design definition in purely digital form; the total information set required to specify and document the product. Related terms include virtual prototyping, virtual product development, soft prototyping, and electronic product development.

direct

When applied to rapid tooling and rapid manufacturing applications, the production of a tool or part from a rapid prototyping device without secondary manufacturing operations.

direct AIM

Injection-mold tooling produced directly from a stereolithography process, where the acronym, AIM, stands for ACES injection molding. *See also ACES*™.

direct digital manufacturing

Application of additive technologies (rapid prototyping) to the production of finished goods without the use of tooling or secondary processes.

direct digital tooling

Application of additive technologies (rapid prototyping) to the creation of molds or dies without the use of secondary or intermediary steps.

direct metal deposition (DMD)

A laser-based technology producing fully dense metal tools. A proprietary rapid tooling process from Precision Optical Manufacturing (POM), it is often applied to tool restoration.

direct metal laser sintering (DMLS)

A rapid prototyping and tooling process from EOS, GmbH that sinters metal powders.

direct shell production casting (DSPC)

A rapid prototyping and tooling process from Soligen based on MIT's 3DP technology. Inkjet deposition of liquid binder onto ceramic powder forms shell molds for investment casting.

DirectTool®

A rapid tooling process from EOS, GmbH for the production of metal tools using the company's direct metal laser sintering technology.

DMLS

See direct metal laser sintering.

drop-on-demand (DOD)

Inkjet methodology now incorporated in rapid prototyping systems, where the material is deposited in a non-continuous stream. Drops are produced and deposited only as required.

DOE

Acronym for design of experiments. A methodology used for running a statistically significant battery of tests (or computer simulations) on a design to determine its sensitivity to, or robustness for, design or manufacturing variations.

DPI

Acronym for dots per inch. Measure or resolution common to computer monitors and also applied to some raster-based rapid prototyping technologies where dots are equated to pixels or a single droplet of material.

DSPC

Acronym for direct-shell production casting. *See direct shell production casting.*

DXF

Acronym for drawing exchange file. A file format that allows for transfer of computer-automated design (CAD) data among dissimilar systems, which was originally devised by Autodesk for the AutoCAD® software program.

E

E_c

Critical energy. A variable for photo-curable materials that specifies the energy required to solidify a given thickness of material. Combined with D_p, the two identify the photo speed of a resin.

EDM

Acronym for electrical discharge machining. Electric current passed through a graphite or copper alloy electrode that machines metal with spark erosion. The electrodes have the same geometry as the intended part or profile to be produced (machined).

early adopters

Customers who, relying on their intuition or vision, buy into new product concepts or new manufacturing processes very early in their life cycles.

economies of scale

Achieving low per-unit costs by producing in volume, permitting fixed costs to be distributed over a large number of products.

economies of scope

Achieving low per-unit costs by computerizing production; allows goods to be manufactured economically in small lot sizes.

electron beam melting

A proprietary rapid prototyping and tooling process from Arcam AB that solidifies metal powder with an electron beam.

element

The basic building block used in geometric modeling. Elements include points, lines, curves, surfaces, and solids.

enterprise prototyping center

Rapid prototyping devices characterized by higher throughput, larger physical size, increased operator control, improved accuracy, and enhanced surface finish. Often operated by a dedicated staff in a lab-like setting.

epoxy tooling

Indirect rapid tooling process where the mold is created by casting an epoxy resin, usually mixed with aluminum powder, against a pattern. The process is suitable for injection molding in low quantities.

ergonomics

Interaction of technological and work situations with the human being. Also called human factors.

extrusion

Process where material, often in a molten or semi-molten state, is forced through an orifice that gives the material shape.

F

facet

Polygonal element that represents the smallest unit of a 3D mesh. These elements can be either three or four sided. The mesh represents an approximation of the actual geometry. Three-sided (triangular) facets are used in stereolithography (STL) files. Both three and four sided elements are used in finite element modeling.

facet deviation

Maximum distance between the triangular element of a stereolithography (STL) file and the surface that it approximates. *See also chord height.*

family mold

Tool that has cavities for two or more different parts.

FDM

Acronym for fused deposition modeling. *See fused deposition modeling.*

FEA

Acronym for finite element analysis. *See finite element analysis.*

feature

Discrete attributes of a model or prototype that include holes, slots, ribs, bosses, snap fits, and other basic elements of a product design.

feature-based modeling

Computer-aided design (CAD) modeling method defined by a series of rules used to describe how features interact with each other to construct a specific solid. Example: The through-hole feature follows the rule that it must pass completely through the part and will do so no matter how the part changes.

finite element analysis

Method used in computer-aided design/computer-aided engineering (CAD/CAE) for determining the structural integrity of a part by mathematical simulation of the part and its loading conditions. Also used to predict the behavior of parts under a thermal load.

first to market

Initial product that creates a new product category.

fixture

Used to hold and position the workpiece for a manufacturing operation.

form and fit

Shape and size of a component and its relationship to mating components. Often used in the context of design analysis to determine the adequacy of a part in terms of its size, shape, and conformance to constraints imposed by mating or nested components.

free-form fabrication

Alternative term for rapid prototyping. Intended to describe a broader base of application where components are generated directly from digital data. *See rapid prototyping.*

free-form surface

Contours that cannot be defined with simple linear or quadratic mathematical equations. Many natural shapes, such as the human face, are examples.

FTP

Acronym for file transfer protocol. A communication standard used for transferring data over the Internet or internal networks.

functional testing

Evaluation of a prototype, in conditions similar to those that the product will experience, to determine its ability to operate as specified.

fused deposition modeling
Rapid prototyping process by Stratasys, Inc. The process extrudes a thermoplastic material and deposits it on a layer-by-layer basis to form a part.

G

GARPA
Acronym for Global Alliance of Rapid Prototyping Associations. An alliance of associations, including the Rapid Prototyping Association of the Society of Manufacturing Engineers (RPA/SME), from around the world that fosters the transfer of information related to rapid prototyping.

gradient material
Graduated displacement of one material with another that yields a gradual transition between two materials.

gross profit
Financial measure that equals sales revenue less variable expense.

I

IGES
Acronym for initial graphic exchange specification. A standard format used for the exchange of 2D and 3D computer-aided design (CAD) data between dissimilar CAD software systems.

indirect
When applied to rapid tooling and rapid manufacturing applications, the production of a tool or part from a rapid prototyping device where secondary manufacturing operations are required between the rapid prototyping operation and the production of the desired item.

injection molding
Manufacturing process where molten plastic is introduced into a tool or die with the use of pressure. It is

commonly applied to prototype and production requirements.

interference checking

Computer-aided design (CAD) capability that automatically examines the intersection of objects within a 3D
model.

investment casting

A manufacturing process that utilizes an expendable
pattern (the investment) to produce metal parts. A mold
is made by repeatedly dipping the pattern in a ceramic
slurry solution followed by fine-grain silica sand. The
pattern is then burned out in an autoclave or furnace,
which simultaneously sinters and strengthens the
ceramic shell. Molten metal is then poured into the
shell. After cooling and solidification, the shell is
destroyed to reveal the final metal part.

K

Keltool®

See 3D Keltool.

kirksite

A low-melting-point alloy often used in the casting of
molds and tools to produce low quantities of parts.

L

Laminated Object Manufacturing®

A patented and trademarked rapid prototyping system,
originally from Helisys, Inc., and now offered by Cubic
Technologies. The system uses a laser to cut a cross-
section from sheet material. The cross-sections are
stacked and bonded together to create an object.

LaserCusing

A term derived from the "concept of laser fusing." A
rapid prototyping and tooling process from Concept
Laser, GmbH, the process produces fully dense metal

parts from powders fused with a high-energy laser.

laser sintering
A rapid prototyping process that uses heat generated by a laser to fuse powdered materials, including plastics and metals.

layer (CAD)
A logical separation of data viewed individually or in combination, which is similar in concept to transparent acetate overlays.

layer (RP)
A thin horizontal slice of the stereolithography (STL) file used to fabricate a rapid prototype, which is typically 0.001–0.010 in. (0.03–0.25 mm) in thickness. *Also see slice.*

layer-based manufacturing
See rapid prototyping or rapid manufacturing.

layer thickness
The vertical dimension of a single slice of a stereolithography (STL) file. Smaller dimensions often lead to smoother surfaces but may increase build time.

LENS
Acronym for laser-engineered net shaping. A rapid prototyping and tooling process that injects metal powder into a pool of molten metal created by a focused laser beam. The process was originally developed at Sandia and later commercialized by Optomec, Inc.

LOM®
See Laminated Object Manufacturing.

LS
See laser sintering.

M

machining
A general term for all manufacturing processes that

produce parts or tools through the removal of material.

manufacturability
The extent to which a product can be easily and effectively manufactured at minimum cost and with maximum reliability.

mass customization
A method of production that stresses the manufacturing of small lots of customized goods rather than large volumes of standardized products.

mass production
Large-scale, high-volume manufacturing of standardized parts that relies on economies of scale to achieve low per-unit costs.

mass properties
The characteristics of a solid, which include volume, weight, center of gravity, and moment of inertia.

MJM
See multi-jet modeling.

mold inserts
Components of a mold core or cavity used to change the geometry features in the mold. Provides alternatives to making multiple molds. Or, it is used in the repair of hardened molds to prevent degradation of the surrounding metal if welding was used for the repair. After being inserted into a frame, a complete core and cavity with ejector mechanism and cooling are then installed into a molding machine.

MRI
Acronym for magnetic resonance imaging.
1) Used medically to scan patients as a non-invasive method to check internal anatomy.
2) Process that uses magnets to align electrons before creating a computer image. The image can be used to generate a 3D file, which is then used to generate a rapid prototype.

3) A technique similar to computer-aided tomography (CT) scanning used to examine internal geometry or structures.

multi-jet modeling
Rapid prototyping processes from 3D Systems that use inkjet technology to deposit materials.

N

NC
Acronym for numerical control. A method of controlling the cutter motion of a machine tool through the use of numerical data and standardized codes. In contrast to computer numerical control (CNC) devices, NC tools offer automation with limited programming ability and logic beyond direct input.

neutral file
A format for electronic data that can be imported and exported by dissimilar software programs. Example file formats include drawing exchange file (DXF), initial graphic exchange specification (IGES), standard for the exchange of product model data (STEP), and stereo-lithography (STL).

NURBS
Acronym for non-uniform rational b-spline. The mathematical description of a surface created by two or more b-splines.

O

OEM
Acronym for original equipment manufacturer. A company that uses product components from one or more vendors to build a product that it sells under its own company name and brand.

outsource

To subcontract services, such as prototyping, design, or manufacturing, to an organization that is independent of the buying (requesting) organization.

P

paper lamination technology (PLT)

A rapid prototyping process from Kira Corporation that laminates paper and then cuts the layer profile with a computerized knife.

parametric CAD

A type of computer-aided design (CAD) methodology that relates the geometry of different elements of a part such that the change of one element changes related features. The association is based on a predetermined correlation.

pattern

A physical representation of a design used to produce molds, dies, or tools. Also called a master pattern.

PDM

Acronym for product data management. A technology used to manage and control all engineering and manufacturing data.

PHAST

Acronym for prototype hard and soft tooling. A proprietary rapid tooling process developed by Procter & Gamble that was granted to the Milwaukee School of Engineering for further process development and refinement.

photopolymer

Liquid resin material that utilizes light (visible or ultraviolet) as a catalyst to initiate polymerization, in which the material cross-links and solidifies. This technique is used by various rapid prototyping technologies.

pipeline management
A process that integrates product strategy, project management, and functional management to continually optimize the cross-project management of all development-related activities.

pixel
An individual dot on a cathode-ray tube (CRT) that, when combined with neighboring dots, creates an image (for example, a television or computer monitor).

plaster mold casting
A process for creating small quantities of metal parts in aluminum, zinc, or magnesium. It is often used as a prototype method for the simulation of die castings. The mold is created from a pattern with several intermediate steps. Metal is cast into the mold, as with investment casting, and the mold is destroyed to yield the metal casting.

PolyJet™
A rapid prototyping process from Objet Geometries that deposits photo-curable materials through an inkjet process.

post-processing
A common practice that includes clean-up and finishing procedures on models after they are removed from the rapid prototyping machine. It may also include mechanical or chemical removal of support structures, powder removal, and surface finishing.

preproduction unit
A product that looks and works like the intended final product, but is made either by hand or in pilot facilities rather than by the final production process.

primitive
1) The lowest state of a solid model.
2) A solid entity that is not derived from other elements, such as a cube, cone, cylinder, or sphere.

product data

All engineering data necessary to define the geometry, function, and behavior of a product over its entire life span. This includes logistic elements for quality, reliability, maintainability, topology, relationship, tolerances, attributes, and features necessary to define the item completely for the purpose of design, analysis, manufacture, test, and inspection.

production tooling

1) Hardened tooling intended to create large volumes (quantities) of parts. The molds should last the life of the products produced. Typically machined from steel, they are used for the mass production of wax, polymer, or metal components.

ProMetal™

A rapid prototyping and tooling process commercialized by Extrude Hone, Inc., which is based on the Massachusetts Institute of Technology (MIT) 3DP technology. The process generates a "green" part by solidifying metal powder with a binder. The green part is placed in a furnace to burn off the binder, sinter the powder, and infiltrate with an alloy.

prototype

A physical model of a part or product made during the product development process. Depending upon the purpose, prototypes may be non-working, functionally working, or functionally and aesthetically complete. Derived from the Latin term for "first form."

prototype tooling

Short-life molds and dies used in the fabrication of molded, stamped, or cast parts. This approach has a limited life expectancy as compared to hardened production tooling. It may yield from one to as many as 50,000 parts, depending on the methods and materials utilized.

Q

QuickCast™

A trademarked process of 3D Systems for a stereolithography build style that reduces the mass of the pattern to accommodate the investment casting process.

R

rapid manufacturing

Production of end-use parts—directly or indirectly—from a rapid prototyping technology.

rapid prototyping

A collection of technologies driven by computer-aided design (CAD) data. They produce physical models and parts through an additive process.

rapid tooling

Production of tools, molds, or dies—directly or indirectly—from a rapid prototyping technology.

raster

1) A two-dimensional array of pixels which, when displayed, form an image or representation of an original document.

2) A scan pattern (as of the electron beam in a cathode-ray tube) produced by scanning an area from side to side in lines from top to bottom. Antonym: vector.

reaction injection molding

A manufacturing process where thermoset resins are injected into rigid tools.

redlining

1) Facility for annotating on-screen documents that allows the transmittal of overlaid comments and sketches.

2) Process of marking documentation for requested changes to a part, tooling, or specification.

rendering
The process of adding shading, color, reflectivity, texture, and other visual elements to a solid model to make it appear realistic.

resin
A general classification of non-metallic materials and compounds. For rapid prototyping, the term is most often associated with the liquid state of stereolithography photopolymers. For molding operations, the term is a reference to any thermoplastic or thermoset material.

return on investment
A financial calculation that illustrates the value of an investment in a specific period of time: [(financial gain − cost)/cost] × 100%.

reverse engineering
Process for the capture of the geometric definition of a physical part through scanning technologies. The resulting data, often a set of spatially oriented discrete points, are imported into a computer-aided design (CAD) system and used for further product refinement, prototype creation, tooling creation, or manufacturing.

RFP
Acronym for request for proposal. A bid package submitted to potential vendors to solicit price and delivery information for a program or project.

RFQ
Acronym for request for quotation. Similar to a request for proposal (RFP), it is generally used when requesting individual parts.

RIM
See reaction injection molding.

road
Term applied to the fused deposition modeling process that describes the extrusion of material in a single pass.

ROI

See return on investment.

RP

See rapid prototyping.

RPA/SME

Acronym for the Rapid Prototyping Association of the Society of Manufacturing Engineers. The association is dedicated to the collection and sharing of information on rapid prototyping, tooling, and manufacturing.

rp-ml

Abbreviation for rapid prototyping mailing list. An Internet forum for the online discussion of topics related to rapid prototyping.

RTV molding

See silicone rubber molding.

rubber molding

See silicone rubber molding.

rubber plaster molding

See plaster mold casting.

S

sandcasting

A manufacturing process for the production of metal castings, including those of gray iron. Sand is packed against a form (tool) to create each half of the tool. After combining the tool halves, metal is cast into the cavity and allowed to cool. To remove the metal casting, the sand tool is destroyed.

selective laser melting

A rapid prototyping and tooling process from F&S, GmbH that produces 100% dense metal parts by melting powdered metal with an infrared laser.

selective laser sintering

A rapid prototyping process, originally developed by DTM Corporation and now owned by 3D Systems, which uses a CO_2 laser to fuse powdered materials, including plastics and metals.

service bureau

1) A company or group of companies providing engineering, prototyping, or manufacturing support to other companies who do not have these capabilities.

2) A commercial entity specializing in providing rapid prototyping and peripheral services to a customer base.

SGC

See solid ground curing.

short-run tooling

Molds created for low-volume production (for example, less than 100 samples).

silicone rubber tooling

Soft tooling created by utilizing room-temperature vulcanized (RTV) rubber material to form molds cast from machined or rapid prototype patterns. Commonly used to produce small lots (25–100 pieces) in urethane materials.

sinter

Heating a material to a temperature below its melting point to fuse it, creating a solid mass.

SL

See stereolithography.

SLA®

Acronym for stereolithography apparatus. A trademarked acronym of 3D Systems, it refers to the machines that use the stereolithography process.

slice

A single layer of a stereolithography (STL) file that becomes the working surface for the additive process.

SLM

See selective laser melting.

SLS®

See selective laser sintering.

solid free-form fabrication
An alternative term for rapid prototyping.

solid ground curing
A rapid prototyping process that solidifies photo-curable materials through a photo mask. The use of the mask allows curing of a complete layer with one flash of ultraviolet (UV) light. This process is no longer available.

solid imaging
An alternative term for rapid prototyping.

solid modeling
A 3D computer-aided design (CAD) technique that represents all the physical characteristics of an object, including: volume, mass, and weight.

solid object

ultraviolet laser printer
A stereolithography process offered by CMET.

SOUP
Acronym for solid-object ultraviolet laser printer. *See solid-object ultraviolet laser printer.*

spin casting
A process that uses rubber molds to create metal castings in low-melting-temperature alloys. The mold is rotated and material is poured into its center. Centrifugal force coats the mold with molten material.

Sprayform™
A trade name and technology owned by the Ford Motor Company. This process uses a wire arc spray of metal alloy onto a ceramic mold pattern to generate tooling.

spray metal tooling
A process for creating prototype or bridge tooling through metal deposition onto a pattern using wire arc spray, vacuum plasma deposition, or similar technique. After creation of the metal tool face, epoxy or another material is used to backfill the tool to add strength. Often used for injection molds.

SRP

Acronym for subtractive rapid prototyping. A process developed by Roland DG. It is used to identify rapid prototyping devices that remove material for prototype creation. Antonym: additive rapid prototyping (ARP).

stair stepping

A result of additive processes where surfaces that are neither vertical nor horizontal are not smooth, since they are approximated by individual layers.

STEP

Acronym for standard for the exchange of product model data. A file format standard for the transfer of data between dissimilar computer-aided design (CAD) systems. The standard was adopted by the International Organization for Standardization (ISO) in December 1994.

stereolithography

A process that builds an object, a layer at a time, by curing photosensitive resin with a laser-generated beam of ultraviolet radiation. Originally applied to the 3D Systems technology, the use of the term has broadened to include all technologies that process prototypes in this manner.

STL

Acronym for stereolithography. A neutral file format exported from computer-aided design (CAD) systems for use as input to rapid prototyping equipment. The file contains point data for the vertices of the triangular facets that combine to approximate the shape of an object.

support structure

Common to many rapid prototyping processes, a scaffold of sacrificial material upon which overhanging geometry is built. Also used to rigidly attach the prototype to the platform upon which it is built. After prototype construction, it is removed in a post-processing operation.

surface

The boundary defining an exterior or interior face of a 3D computer-aided design (CAD) model.

surface modeling

See surface wireframe.

surface normal

A vector perpendicular to a surface or facet in a stereo-lithography (STL) file. In the STL file, the direction of the vector indicates the outward facing side of the facet.

surfaced wireframe

A method of 3D computer-aided design (CAD) modeling that represents part geometry with bounding edges and skins that stretch between the boundaries. The CAD model is defined by its innermost and outermost boundaries and does not contain any mass between these boundaries.

T

TALC

Acronym for technology adoption life cycle. A business model describing the adoption of technology through an analysis of purchasing traits.

thermoplastic

Plastic compound processed (molded) in a liquid state, which is achieved by elevated temperature. This class of plastics can be repeatedly cycled through the liquid and solid states. Common applications are injection molding, blow molding, and vacuum forming.

thermoset

Plastic compound processed in a liquid state where two or more liquid components are blended just prior to molding. Upon blending, an exothermic chemical reaction causes the liquid to change to a solid state. Unlike thermoplastics, once solidified these materials cannot be returned to a liquid state. Common applications are rubber molding and reaction injection molding.

time to market

Period required to conceive, develop, manufacture, and deliver a new product.

tooling

Generic term used to describe molds or dies used in the production of parts and assemblies. Examples include injection molds, blow molds, die-cast dies, and stamping dies.

U

ultrasonic consolidation

A proprietary rapid prototyping and tooling process from Solidica, Inc. that ultrasonically welds sheet metal to deliver homogeneous material properties. After welding of the sheet material, the profile is computer numerical control (CNC) machined.

urethane

A thermoset material commonly used in rubber molding and reaction injection molding processes. Any of various polymers that contain NHCOO linkages.

UV

Acronym for ultraviolet. Light energy situated beyond the visible spectrum at its violet end, having a wave-length shorter than the wavelengths of visible light and longer than those of X-rays. Often used in the curing of photopolymer resins.

V

vacuum forming

A process for producing plastic parts by heating plastic sheet and drawing it against a form when air is pulled through the form.

vector
A quantity that has magnitude and direction. It is
commonly represented by a directed line segment
whose length represents the magnitude and whose
orientation in space represents the direction.

virtual prototyping
Computer-based generation of 3D geometry for
analyzing product design features where the digital data
is presented with realism. Often associated with immer-
sive environments that offer an ability to interact with
the digital design as if it were real. More commonly
applied to computer-based testing and analysis methods
such as finite element analysis.

voxel
Short term for "volume cell." The 3D equivalent of the
pixel.

W

WaterWorks™
A trademarked and patented process of Stratasys, Inc.
used with the fused deposition modeling process,
which allows models (or assemblies) to be made with
movable parts already assembled. The process dissolves
support material in a water-based solution.

wireframe
A computer-aided design (CAD) modeling method that
defines a part by its innermost and outermost boundaries. The model does not contain any mass between
the boundaries or any bounding surfaces.

Index

W

water-soluble support structure (WaterWork™), 201, 203, 233, 261
websites, 351, 362–363

Z

Z axis, 60–62, 69–70
Z Corporation technology, 20, 87, 98, 163, 275